# Sociocultural Perspectives on Volatile Solvent Use

ALL HARRINGTON PARK PRESS BOOKS
ARE PRINTED ON CERTIFIED
ACID-FREE PAPER

# Sociocultural Perspectives on Volatile Solvent Use

Fred Beauvais
Joseph E. Trimble
Editors

*Sociocultural Perspectives on Volatile Solvent Use,* edited by Fred Beauvais and Joseph E. Trimble, was simultaneously issued by The Haworth Press, Inc., under the same title, as a special issue of *Drugs & Society,* Volume 10, Numbers 1/2 1997, Bernard Segal, Editor.

Harrington Park Press
An Imprint of
The Haworth Press, Inc.
New York • London

ISBN 1-56023-096-7

Published by

Harrington Park Press, 10 Alice Street, Binghamton, NY 13904-1580 USA

Harrington Park Press is an imprint of The Haworth Press, Inc., 10 Alice Street, Binghamton, NY 13904-1580 USA.

*Sociocultural Perspectives on Volatile Solvent Use* has also been published as *Drugs & Society*, Volume 10, Numbers 1/2 1997.



Cover design by Monica Seifert

**Library of Congress Cataloging-in-Publication Data**

Sociocultural perspectives on volatile solvent use / Fred Beauvais, Joseph E. Trimble, editors.
p. cm.
Includes bibliographical references and index.
ISBN 1-56023-096-7 (alk. paper)
1. Solvent abuse. I. Beauvais, Fred. II. Trimble, Joseph E.
HV5822.S65S62 1997
362.29'9–dc21

97-699
CIP

# INDEXING & ABSTRACTING

Contributions to this publication are selectively indexed or abstracted in print, electronic, online, or CD-ROM version(s) of the reference tools and information services listed below. This list is current as of the copyright date of this publication. See the end of this section for additional notes.

- ***Abstracts in Anthropology,*** Baywood Publishing Company, 26 Austin Avenue, P.O. Box 337, Amityville, NY 11701
- ***Academic Abstracts/CD-ROM,*** EBSCO Publishing Editorial Department, P. O. Box 590, Ipswich, MA 01938-0590
- ***ADDICTION ABSTRACTS,*** National Addiction Centre, 4 Windsor Walk, London SE5 8AF, England
- ***ALCONLINE Database,*** Swedish Council for Information on Alcohol and Other Drugs, Box 27302, S-102 54 Stockholm, Sweden
- ***Applied Social Sciences Index & Abstracts (ASSIA) (Online: ASSI via Data-Star) (CDRom: ASSIA Plus),*** Bowker-Saur Limited, Maypole House, Maypole Road, East Grinstead, West Sussex, RH19 1HH, England
- ***Brown University Digest of Addiction Theory and Application, The (DATA Newsletter),*** Project Cork Institute, Dartmouth Medical School, 14 South Main Street, Suite 2F, Hanover, NH 03755-2015
- ***Cambridge Scientific Abstracts,*** *Health & Safety Science Abstracts,* Environmental Routenet (accessed via INTERNET), 7200 Wisconsin Avenue #601, Bethesda, MD 20814
- ***Child Development Abstracts & Bibliography,*** University of Kansas, 2 Bailey Hall, Lawrence, KS 66045
- ***CNPIEC Reference Guide: Chinese National Directory of Foreign Periodicals,*** P.O. Box 88, Beijing, People's Republic of China
- ***Criminal Justice Abstracts,*** Willow Tree Press, 15 Washington Street, 4th Floor, Newark, NJ 07102

(continued)

- ***Criminal Justice Periodical Index,*** University Microfilms, Inc., 300 North Zeeb Road, Ann Arbor, MI 48106
- ***Excerpta Medica/Secondary Publishing Division,*** Elsevier Science Inc., Secondary Publishing Division, 655 Avenue of the Americas, New York, NY 10010
- ***Family Studies Database (online and CD/ROM),*** National Information Services Corporation, 306 East Baltimore Pike, 2nd Floor, Media, PA 19063
- ***Health Source: Indexing & Abstracting of 160 selected health related journals, updated monthly:*** EBSCO Publishing, 83 Pine Street, Peabody, MA 01960
- ***Health Source Plus: expanded version of "Health Source" to be released shortly:*** EBSCO Publishing, 83 Pine Street, Peabody, MA 01960
- ***Human Resources Abstracts (HRA),*** Sage Publications, Inc., 2455 Teller Road, Newbury Park, CA 91320
- ***IBZ International Bibliography of Periodical Literature,*** Zeller Verlag GmbH & Co., P.O.B. 1949, d-49009 Osnabruck, Germany
- ***Index to Periodical Articles Related to Law,*** University of Texas, 727 East 26th Street, Austin, TX 78705
- ***International Pharmaceutical Abstracts,*** ASHP, 7272 Wisconsin Avenue, Bethesda, MD 20814
- ***International Political Science Abstracts,*** 27 Rue Saint-Guillaume, F-75337 Paris, Cedex 07, France
- ***INTERNET ACCESS (& additional networks) Bulletin Board for Libraries ("BUBL"), coverage of information resources on INTERNET, JANET, and other networks.***
  - JANET X.29: UK.AC.BATH.BUBL or 00006012101300
  - TELNET: BUBL.BATH.AC.UK or 138.38.32.45 login 'bubl'
  - Gopher: BUBL.BATH.AC.UK (138.32.32.45). Port 7070
  - World Wide Web: http: / / www.bubl.bath.ac.uk./BUBL/ home.html
  - NISSWAIS: telnetniss.ac.uk (for the NISS gateway)

  The Andersonian Library, Curran Building, 101 St. James Road, Glasgow G4 ONS, Scotland

(continued)

- ***Medication Use STudies (MUST) DATABASE,*** The University of Mississippi, School of Pharmacy, University, MS 38677
- ***Mental Health Abstracts (online through DIALOG),*** IFI/Plenum Data Company, 3202 Kirkwood Highway, Wilmington, DE 19808
- ***NIAAA Alcohol and Alcohol Problems Science Database (ETOH),*** National Institute on Alcohol Abuse and Alcoholism, 1400 Eye Street NW, Suite 600, Washington, DC 20005
- ***Personnel Management Abstracts,*** 704 Island Lake Road, Chelsea, MI 48118
- ***Psychological Abstracts (PsycINFO),*** American Psychological Association, P.O. Box 91600, Washington, DC 20090-1600
- ***Public Affairs Information Bulletin (PAIS),*** Public Affairs Information Service, Inc., 521 West 43rd Street, New York, NY 10036-4396
- ***Referativnyi Zhurnal (Abstracts Journal of the Institute of Scientific Information of the Republic of Russia),*** The Institute of Scientific Information, Baltijskaja ul., 14, Moscow A-219, Republic of Russia
- ***Sage Family Studies Abstracts (SFSA),*** Sage Publications, Inc., 2455 Teller Road, Newbury Park, CA 91320
- ***Social Planning/Policy & Development Abstracts (SOPODA),*** Sociological Abstracts, Inc., P.O. Box 22206, San Diego, CA 92192-0206
- ***Social Work Abstracts,*** National Association of Social Workers, 750 First Street NW, 8th Floor, Washington, DC 20002
- ***Sociological Abstracts (SA),*** Sociological Abstracts, Inc., P.O. Box 22206, San Diego, CA 92192-0206
- ***SOMED (social medicine) Database,*** Landes Institut fur Den Offentlichen Gesundheitsdienst NRW, Postfach 20 10 12, D-33548 Bielefeld, Germany
- ***Sport Database/Discus,*** Sport Information Resource Center, 1600 James Naismith Drive, Suite 107, Gloucester, Ontario K1B 5N4, Canada
- ***Violence and Abuse Abstracts: A Review of Current Literature on Interpersonal Violence (VAA),*** Sage Publications, Inc., 2455 Teller Road, Newbury Park, CA 91320

(continued)

# SPECIAL BIBLIOGRAPHIC NOTES

*related to special journal issues (separates) and indexing/abstracting*

- ❑ indexing/abstracting services in this list will also cover material in any "separate" that is co-published simultaneously with Haworth's special thematic journal issue or DocuSerial. Indexing/abstracting usually covers material at the article/chapter level.
- ❑ monographic co-editions are intended for either non-subscribers or libraries which intend to purchase a second copy for their circulating collections.
- ❑ monographic co-editions are reported to all jobbers/wholesalers/approval plans. The source journal is listed as the "series" to assist the prevention of duplicate purchasing in the same manner utilized for books-in-series.
- ❑ to facilitate user/access services all indexing/abstracting services are encouraged to utilize the co-indexing entry note indicated at the bottom of the first page of each article/chapter/contribution.
- ❑ this is intended to assist a library user of any reference tool (whether print, electronic, online, or CD-ROM) to locate the monographic version if the library has purchased this version but not a subscription to the source journal.
- ❑ individual articles/chapters in any Haworth publication are also available through the Haworth Document Delivery Services (HDDS).

# *CONTENTS*

## ABOUT THE EDITORS

**Fred Beauvais, PhD,** is a Senior Research Scientist at the Tri-Ethnic Center for Prevention Research within the Psychology Department at Colorado State University in Fort Collins, Colorado. He is also an affiliate faculty member in the Psychology Department at Colorado State University and at the University of Alaska, Anchorage. Dr. Beauvais is on the National Advisory Council for the National Institute on Alcohol Abuse and Alcoholism, serves on the advisory board for the International Institute on Inhalant Abuse, and is on the Scientific Advisory Committee for the American Indian and Alaska Native Mental Health Research Center.

**Joseph E. Trimble, PhD,** is a Professor of Psychology at Western Washington University in Bellingham, Washington. He is a Visiting Senior Scholar at the Tri-Ethnic Center for Prevention Research at Colorado State University. Additionally, Dr. Trimble has held offices in the International Association for Cross-Cultural Psychology and the American Psychological Association. He holds Fellow status in both organizations. In 1994, he received a Lifetime distinguished Career Award from the American Psychological Association's Division 45 for his research and dedication to cross-cultural and ethnic psychology.

# Preface

Volatile solvent abuse constitutes a major problem in the United States and among other nations in the world.[1] Recent estimates, for example, place the number of solvent abusers in Central and South America at nearly 20 million individuals, mostly street children. These are staggering numbers! While the social and economic conditions that spawned this epidemic are not likely to be replicated elsewhere, the numbers do illustrate the potential for solvent abuse. Although the magnitude of the problem in the U. S. is nowhere near this size, the volatile solvent abuse is a growing problem. Results from the Monitoring the Future study show that lifetime prevalence of solvent use among high school seniors has increased steadily from 10.3% in 1976 to 17.4% in 1995 (All drug use continues upward, 1995). During this time period other drug use in the U. S. declined, leading to the conclusion that solvent abuse does not follow the same dynamics as other drugs. Whereas drug prevention programs, or other societal factors, are operating to reduce the experimentation with a majority of drugs, solvent abuse appears to run its own independent course.

The problem of solvent abuse receives comparatively little attention from the treatment, prevention, and research communities. Its existence as an "orphan drug" may account for the increase in use; drug treatment providers and clinical practitioners, as well as the general community, are perplexed by the use of solvents and thus prefer to target other drugs for prevention and treatment. Solvent abusers tend to be markedly different from other drug users. Those who persist beyond the usual experimentation stage of solvent use are dysfunctional in a variety of ways and defy conventional treatment and prevention efforts (Oetting & Webb, 1992; Jumper-Thurman & Beauvais, 1992). Treatment professionals report that they are uniformly unsuccessful with solvent abusers and are reluctant to admit them to their pro-

---

1. See Beauvais (1987) for a discussion of the terms "inhalants" and "volatile solvents."

[Haworth co-indexing entry note]: "Preface." Beauvais, Fred, and Joseph E. Trimble. Co-published simultaneously in *Drugs & Society* (The Haworth Press, Inc.) Vol. 10, No. 1/2, 1997, pp. xi-xii; and: *Sociocultural Perspectives on Volatile Solvent Use* (ed: Fred Beauvais, and Joseph E. Trimble) Harrington Park Press, an imprint of The Haworth Press, Inc., 1997, pp. xi-xii. Single or multiple copies of this article are available for a fee from The Haworth Document Delivery Service [1-800-342-9678, 9:00 a.m. - 5:00 p.m. (EST). E-mail address: getinfo@haworth.com].

 

grams. While it has been proposed that separate treatment programs are necessary for those who abuse solvents, there are only a handful of such programs in the United States. Workers in the area of prevention are equally baffled by what approaches to use to prevent young people from experimenting with or using solvents. Solvents are contained in literally hundreds of household and commercial products and prevention messages run the risk of introducing young people to trying chemicals that they may not otherwise consider. This is a particular problem because the typical age for solvent experimentation is much younger than for other drugs; prevention thus has to take place during a time when young people are very suggestible and naive about the effects of drugs.

The peculiar nature of solvent use places it outside the normal range of substance abuse research, thus our knowledge of the phenomena is quite limited, despite the potential for a great deal of harm from the use of these substances. In recognition of the need for an increase in such knowledge, the Tri-Ethnic Center for Prevention Research at Colorado State University convened a number of researchers who are interested in volatile solvent research to discuss various aspects of the problem. The articles in this volume resulted from this discussion. The content of these articles is quite diverse, and it is hoped that they will kindle some ideas and interest among researchers who might, in the future, devote time and energy to explore the many facets of this most enigmatic form of drug-taking behavior.

There is quite clearly a need for continued inquiry into the phenomena of solvent use and abuse. Each of the articles presented contain parts of that research agenda, with the final chapter reviewing some of the more pertinent issues.

*Fred Beauvais, PhD*
*Joseph E. Trimble, PhD*

## REFERENCES

All drug use continues upward. (1995). *Drugs and Drug Abuse Education Newsletter, 26,* 93-102.

Beauvais, F. (1987). Toward a clear definition of inhalant abuse. *The International Journal of the Addictions, 22,* 779-784.

Jumper-Thurman, P., and Beauvais, F. (1992). Treatment of volatile solvent abusers. In C. Sharp, Beauvais, F. and Spence, R. (Eds.) *Inhalant abuse: A volatile research agenda.* (Monograph series # 109). Rockville, MD: National Institute on Drug Abuse, 203-214.

Oetting, E. and Webb, J. (1992). Psychosocial characteristics and their links with inhalants: A research agenda. *Inhalant abuse: A volatile research agenda* (Monograph Series #129). Rockville, MD: National Institute on Drug Abuse, 59-98.

# Introduction

*"Um. You do it to feel good and be cool, and once you do the drugs you get addicted"*[1]

These words by a Navajo youth describe the situation, not only for inhalants or solvents, but for most all substances that can be tried by youth. Unfortunately, solvent and inhalant abuse constitutes a major problem among American Indian and Alaskan Native youth, one that perplexes elders, parents, and tribal members, as well as professionals working to resolve this problem. Yet, inhalant/solvent abuse affects all youth, and it is a problem, as noted earlier, that is beginning to reach epidemic proportions.

While there are common motives and patterns of drug-taking behavior among youth and young adults, such as using marijuana or drinking to achieve a new and exciting experience, there are also specific factors that contribute to unique patterns of use. Such factors as culture, ethnicity, gender and educational level (e.g., attending vs. dropping-out as risk factors) are all related to initiation to drugs and patterns of use.

The papers in this volume provide a perspective on how such factors are especially related to inhalant or solvent abuse among American Indian and Alaskan Native youth, presenting information that has not been previously cited. All the authors are intensely involved with assisting Native American communities to deal with the effects of drug abuse, particularly inhalant/solvent abuse. Much of what is presented has been learned from working directly in communities affected by the problem. It is anticipated that although the discussion revolves around Native American communities, the findings and conclusions can be generalized to both the larger population and to other ethnic groups within the larger culture.

*Fred Beauvais, PhD*
*Joseph E. Trimble, PhD*

---

1. Navajo youth, speaking about inhalants, cited in Trotter, Rolf and Baldwin, this volume.

[Haworth co-indexing entry note]: "Introduction." Beauvais, Fred, and Joseph E. Trimble. Co-published simultaneously in *Drugs & Society* (The Haworth Press, Inc.) Vol. 10, No. 1/2, 1997, p. 1; and: *Sociocultural Perspectives on Volatile Solvent Use* (ed: Fred Beauvais, and Joseph E. Trimble) Harrington Park Press, an imprint of The Haworth Press, Inc., 1997, p. 1. Single or multiple copies of this article are available for a fee from The Haworth Document Delivery Service [1-800-342-9678, 9:00 a.m. - 5:00 p.m. (EST). E-mail address: getinfo@haworth.com].

# The Three Common Behavioral Patterns of Inhalant/Solvent Abuse: Selected Findings and Research Issues

Philip A. May, PhD
Ann M. Del Vecchio, PhD

**SUMMARY.** One of the necessary steps in understanding behavior is to adequately classify the existing patterns associated with that behavior. The literature contains evidence for at least three common subtypes of inhalant users: (a) young inhalant users, (b) adolescent polydrug users who frequently use inhalants and (c) adult users. Several national and regional data sources are examined for the presence of these types and the categorization is generally upheld. *[Article copies available for a fee from The Haworth Document Delivery Service: 1-800-342-9678. E-mail address: getinfo@haworth.com]*

Philip A. May is Professor of Sociology and Psychiatry, and Director, Center on Alcoholism, Substance Abuse, and Addictions, and Ann M. Del Vecchio is Research Scientist, Center on Alcoholism, Substance Abuse, and Addictions, both at the University of New Mexico.

The authors extend special thanks to Virginia Rood.

Partial clerical support was provided by Grant #T34-MH19101. Some data presented here were collected and analyzed under contract with the Native American Adolescent Injury Prevention Program of the State of New Mexico, Department of Health. Other data originate from the Voices of Indian Teens Project, a joint venture of the University of Colorado Health Sciences Center, National Center for American Indian and Alaska Native Mental Health Research and the University of New Mexico Center on Alcoholism, Substance Abuse, and Addictions. Funding for the latter was provided by NIAAA (RO1AA08474).

[Haworth co-indexing entry note]: "The Three Common Behavioral Patterns of Inhalant/Solvent Abuse: Selected Findings and Research Issues." May, Philip A., and Ann M. Del Vecchio. Co-published simultaneously in *Drugs & Society* (The Haworth Press, Inc.) Vol. 10, No. 1/2, 1997, pp. 3-37; and: *Sociocultural Perspectives on Volatile Solvent Use* (ed: Fred Beauvais, and Joseph E. Trimble) Harrington Park Press, an imprint of The Haworth Press, Inc., 1997, pp. 3-37. Single or multiple copies of this article are available for a fee from The Haworth Document Delivery Service [1-800-342-9678, 9:00 a.m. - 5:00 p.m. (EST). E-mail address: getinfo@haworth.com].

 

The voluntary inhalation of various household and industrial solvents (e.g., glue, gasoline, cleaning fluids, and others) has been practiced for many years. It is only during the past four decades, however, that considerable attention has been given to the problem in medical and behavioral literature. This paper is concerned with the use of volatile household and commercial solvents, exclusive of amyl and butyl nitrites, commonly referred to as nitrites. This article is not concerned with an extensive review of the literature on inhalant abuse and the use of solvents in general; the reader is referred to a number of articles that provide such reviews or present a substantive overview of the problem (c.f., Beauvais, 1992b; Cohen, 1977; Novak, 1980; Remington & Hoffman, 1984; Watson, 1980), and to three monographs published by the National Institute on Drug Abuse (NIDA) (Crider & Rouse, 1988; Sharp, Beauvais, & Spence, 1992; Sharp & Brehm, 1977). What this article is concerned with is a classification of different patterns of inhalant or solvent abuse.

The literature on inhalants and solvents has grown extensively and more detailed in recent years, and the various characteristics of inhalant abuse are generally well documented from a variety of studies. Most of the studies are survey research, but there are others that detail findings from people treated for solvent abuse.

There are, however, gaps in the literature regarding various behavioral issues. This paper will attempt to highlight some of those gaps, particularly concerning behavioral subtypes found among inhalant abusers in the United States.

## *NATIONAL PREVALENCE ACROSS AGE GROUPS*

National prevalence data are presented in Tables 1 and 2. The data on inhalant use in Table 1 (from the National Institute on Drug Abuse, National Household Survey, 1994) reveal that the highest lifetime prevalence ("ever used") is recorded among males in the 18-25 and 26-34 age categories. Approximately 12% of all males in these age groups report having used inhalants, and 7.4% to 6.1% of the females in these age categories report use. However, when the rates of use in the "past year" are examined, a different picture emerges. Approximately 4% of all males and females in ages 12-17 have used inhalants in the past year, which is only slightly higher than the use rates in ages 18-25. Use in the "past month" shows that a lower percentage–approximately 2%–of males and females between the ages of 12-17 have used inhalants in the past month. Inhalant use tapers off in the late teens and twenties and is reduced to virtually nothing by ages 35 and above.

TABLE 1. Inhalants: Ever, Past Year, and Past Month (1991) by Sex and Age Groups for Total U.S. Population

| Age | Ever Used | Used Past Year | Used Past Month |
|---|---|---|---|
| | Rate Estimates | | |
| | Observed Estimates (%) | Observed Estimates (%) | Observed Estimates (%) |
| 12-17 | 7.0 | 4.1 | 1.8 |
| Male | 7.1 | 4.2 | 1.6 |
| Female | 7.0 | 3.9 | 2.0 |
| 18-25 | 10.9 | 3.5 | 1.5 |
| Male | 12.3 | 4.6 | 2.0 |
| Female | 9.6 | 2.4 | 1.1 |
| 26-34 | 9.2 | 0.9 | 0.5 |
| Male | 12.2 | 1.3 | 0.6 |
| Female | 6.4 | 0.6 | 0.4 |
| 35+ | 2.7 | 0.6 | 0.2 |
| Male | 3.9 | 0.7 | 0.3 |
| Female | 1.7 | 0.6 | 0.1 |
| TOTAL | 5.6 | 1.4 | 0.6 |
| Male | 7.1 | 1.7 | 0.7 |
| Female | 4.1 | 1.1 | 0.5 |

Source: NIDA, National Household Survey on Drug Abuse, 1991.

TABLE 2. Trends in Prevalence of Inhalants for Five U.S. Populations: 8th, 10th, and 12th Graders, College Students, and Young Adults 19-28 Years Old

| | Lifetime | | Annual | | 30-Day | | Daily | |
|---|---|---|---|---|---|---|---|---|
| | 1991 | 1993 | 1991 | 1993 | 1991 | 1993 | 1991 | 1993 |
| **Inhalants**[1,2] | | | | | | | | |
| 8th Grade | 17.6 | 19.4 | 9.0 | 11.0 | 4.4 | 5.4 | 0.2 | 0.3 |
| 10th Grade | 15.7 | 17.5 | 7.1 | 8.4 | 2.7 | 3.3 | 0.1 | 0.2 |
| 12th Grade | 17.6 | 17.4 | 6.6 | 7.0 | 2.4 | 2.5 | 0.2 | 0.1 |
| College Students | 14.4 | 14.8 | 3.5 | 3.8 | 0.9 | 1.3 | – | – |
| Young Adults | 13.4 | 14.1 | 2.0 | 2.1 | 0.5 | 0.7 | * | * |

Source: Johnston, O'Malley, and Bachman, 1994.
(–) indicates that data are not available.
(*) indicates less than 0.05 percent.

[1] 12th grade, college students, and young adults only: Data based on five questionnaire forms in 1991-1993; N for 12th graders is five sixths of N indicated. N for college students is 1250 in 1993, and N for young adults is 5480.

[2] Inhalants are unadjusted for underreporting of amyl and butyl nitrites; hallucinogens are unadjusted for underreporting of PCP.

Table 2 presents the prevalence of inhalant use for students and young adults under the age of 28 years (from the NIDA-funded project, Monitoring the Future, Johnston, O'Malley and Bachman, 1994). A higher use rate is recorded in this survey than in the National Household Survey. In-school surveys for 1993 reveal that eighth graders have the highest lifetime, annual, and 30-day use patterns of any of the students. Overall, about 19% of eighth graders have used inhalants in their lifetimes, 11% in the past year, and 5.4% in the last 30 days. Only 0.3%, however, use inhalants daily. Among college students about 15% say that they have ever used, 4% have used in the past year, and approximately 1% have used in the last 30 days. Young adults report similar use rates for lifetime experience but lower use rates for annual, 30-day, and daily experience. In fact, the daily use in this category is less than 0.05%.

The above data reflect the general population of the United States, but some variance has been found for different ethnic groups. Beauvais (1992b) reports, for example, that American Indians (17%), Spanish Americans (16%), Mexican Americans (12%), and White, non-Hispanics (17%) have higher rates of lifetime prevalence use than African Americans (8%). Furthermore, special studies in various parts of North America have indicated that some Hispanic and American Indian communities have been overcome by fads that produce particularly high prevalence for certain time periods. In Canada for example, Smart (1988) reports that many Canadian Native communities have a significantly higher problem than do non-Native communities in general but, on the other hand, some are not affected by waves of popularity for inhalant use and therefore have low rates. Remington and Hoffman (1984) and Lalinec-Michaud, Subak, Ghadirian, and Kovess (1991) independently reported on high-risk Native communities in Canada, where inhalant abuse is very high. Liban and Smart (1982), however, found no difference between Canadian Natives and non-Natives in middle-class, urban environments. Similar trends and variability have been found among American Indian groups in the United States. For a review of the American Indian literature, one might consult Beauvais, Oetting and Edwards (1985), Beauvais (1992a), or May (1982).

Hispanic American studies, which have found substantially higher rates of inhalant use, were first published in the 1960s (see De la Rosa, Khalsa, & Rouser, 1990; and E. R. Padilla, A. M. Padilla, Morales, Olmedo, & Ramirez, 1979), but recent results show extreme variation by community (Mata & Andrew, 1988). Chavez and Swaim (1992) report that lifetime prevalence of use is no higher nationally among Hispanics than among non-Hispanic whites, but that current use of inhalants in the past month among Hispanic eighth graders is indeed higher.

To summarize this prevalence literature, inhalant abuse tends to be highest among young people, reaching its peak at about age 13 (Beauvais, 1992b). Although there is variance among ethnic groups in a number of studies, the variance within ethnic groups is also great from one community to the next. In fact, influences such as culture, urbanity, local community norms, and fads greatly influence ethnic-group participation in inhalant/solvent use (Oetting & Webb, 1992).

## *TRAITS OF INHALANT/SOLVENT ABUSERS*

Oetting and Webb (1992) reviewed the psychosocial characteristics of inhalant abusers and pinpointed a number of variables that consistently appeared in the literature. Two of the more important findings are: (a) the families of inhalant abusers are generally characterized by serious dysfunction of various kinds (Oetting & Webb, 1992); and (b) family discord, aggression, and hostility are frequently found in families of drug users (Babst, Deren, Schmidler, Lipton, & Dembo, 1978) and particularly in families of inhalant/solvent abusers (see, for example, Korman, Trumboli, & Semler, 1980).

Furthermore, there is generally a substantial history of broken homes and high use of alcohol and drugs among siblings and parents (Albaugh & Albaugh, 1979; De la Rosa et al., 1990; Korman et al., 1980; Reed & May, 1984). Families of inhalant abusers were most frequently found to be of lower socioeconomic status (Oetting & Webb, 1992; Reed & May, 1984). Some authors have not found any socioeconomic status differences between inhalant abusers and other groups, but Oetting and Webb link low social status to an opportunity for access to inhalant use and abuse. Indeed, in some families, cultures, and communities, youths may have more siblings, extended family, and peer group members from whom to learn the values and techniques of drug use patterns (Swaim, Oetting, Thurman, Beauvais, & Edwards, 1993).

Crime has been found to be frequent among inhalant abusers. Although some studies have linked violent behavior to inhalant abuse (Simons & Kashani, 1980; Tinklenberg, P. Murphy, P. Murphy, & Pfefferbaum, 1981), others have found that inhalant abusers tend to be much more criminal and deviant in all ways (c.f., Jacobs & Ghodse, 1988; Reed & May, 1984). When compared with other delinquents arrested and processed through a criminal justice unit, inhalant abusers were found to have criminal records 3 times greater than those of randomly selected delinquents, and over 2 times greater than a matched, stratified sample of delinquents (Reed & May, 1984).

Inhalant abusers generally have serious problems in their studies and in school attendance (Wingert & Fifield, 1985). In the review by Oetting and Webb, a number of studies are cited to demonstrate that inhalant abusers are more likely to be absent, to be suspended, to drop out, and to have been expelled (Coulehan et al. 1983; Oetting & Webb, 1992). Furthermore, they are generally portrayed as having lower grades, more demerits, and fewer positive experiences in school (Wingert & Fifield, 1985).

A significant psychological characteristic found among inhalant abusers is intra- and interpersonal difficulties (Oetting & Webb, 1992). Emotional distress, lower self-esteem, and adolescent adjustment reaction also are common among inhalant abusers, and some studies have indicated that personality disorders may also exist. Korman et al. (1980) found that in addition to intra- and interpersonal difficulties among inhalant abusers, they also manifested withdrawal/isolation, suspiciousness/hostility, dissociation/depersonalization, sexual difficulties, marital discord, family discord, school problems, employment problems, and trouble with the law to a greater extent than found among polysubstance abusers or nondependent groups. Inhalant abusers were also reported by Korman et al. (1980) to sleep excessively and, when functioning, they were believed to be a potential threat to themselves and to others.

Depression is also believed to be common among solvent abusers (Jacobs & Ghodse, 1987; Zur & Yule, 1990). Additionally, poor appearance, including poor hygiene and inappropriate dress, and cognitive difficulties, were found to exist among inhalant abusers (Korman et al. 1980).

Oetting and Webb (1992), and others, have indicated that culture plays a part in inhalant abuse. In some communities, inhalant abuse becomes more attractive owing to social fads and group practices. Of particular importance is the fact that most children begin and continue inhalant abuse in peer clusters that support its use. Both those who use inhalants on an isolated basis, and those who are polysubstance abusers, learn and practice most inhalant abuse in these peer group associations (Kauffman, 1973). Oetting and Webb (1992) estimate that more than three-fourths of inhalant use is with friends. It is thus apparent that subculture and cultural influences have a role in the adoption and maintenance of inhalant abuse.

## *UNDERDEVELOPED TOPICS*

A review of the literature reveals several deficiencies pertaining to knowledge about inhalant abuse. First, the specification of exactly which solvents are being used is not well documented or described. Most inhalants and solvents are treated equally, and method of administration is generally taken for granted as inhalation. That is, it is believed that most substances are ingested through sniffing (nose only) or huffing (use of mouth and nose simultaneously). Inhalation is, indeed, the most common mode of administration. There is, however, another use pattern involving inhalants that warrants further investigation. There are, for example, a number of commercial products that combine both volatile solvents (or gaseous substances) and alcohol; hair spray is one example. There are reports that some individuals will turn the hair spray can upside down and collect the propellant (oftentimes propane) in a bag which is then huffed. When the propellant is exhausted the can is then opened and the remaining contents are consumed by drinking; most likely at this stage it is the alcohol content that is being sought. This method thus represents a combined practice involving both inhalants and alcohol, but it is unclear whether those engaging in this behavior are making the distinction between the effects and physical reactions of the two substances. The difference must be recognized at some level because there are virtually no reports of individuals drinking gasoline, for example, to get intoxicated. It is clear that there are some rather unusual practices involving inhalants that need to be examined further.

Another informational gap pertains to adult use. Much of the literature reports on school-based surveys, which results from easy access to younger individuals. But, because students drop out of school, or graduate and make a transition to early adulthood, survey procedures become increasingly difficult to implement. Follow-up studies on adults are thus quite rare (Beauvais, 1992b; Hersey & Miller, 1982; Oetting, Edwards, & Beauvais, 1988). There is thus a lack of information about the progression of inhalants and solvent use among young adults who continue to use inhalants, and little is known about the eventual outcome of these individuals. Research is therefore needed that investigates age-specific patterns and consequences of inhalant abuse.

Another informational gap that remains are data pertaining to differences between behavioral patterns of use. As Oetting et al. (1988) and McSherry (1988) have pointed out, there may be several discrete types of inhalant abusers. Although the literature provides a number of specific psychosocial and behavioral characteristics to examine as risk factors for inhalant abusers, it has seldom indicated which of these risk factors are more important for young users, midcareer users, and users in latter, adult phases of life.

What follows is an examination of some of the evidence that points toward three types or behavioral patterns of inhalant use, initially described by Oetting et al. (1988). Utilizing data from various studies in the southwestern United States and some specific articles in the literature, an attempt is made to further define these behavioral subtypes.

## THREE BEHAVIORAL PATTERNS OF INHALANT ABUSE

McSherry (1988), in an article describing treatment program experiences with solvent abusers in Philadelphia, Pennsylvania, listed three types of solvent abusers. The first, the *experimental abuser,* has under two years of experience with solvents, and use is sporadic. There is little criminal activity and little evidence that the use of other drugs is associated with this pattern of use. The age range is between 14 and 17 years.

The second type McSherry defined is the *acute abuser,* described as one who has been using solvents for two to four years and who uses them at least three times weekly. There may be some involvement with petty criminal activity and with other drugs and alcohol but, according to McSherry, solvents are the predominant drug of choice. The age range of these abusers is between 17 and 21 years.

The third type that McSherry listed is the *chronic abuser.* This individual has been using inhalants and solvents for 5 to 15 years, and solvents are the drug of choice; the individual is psychologically and physically addicted to the use of solvents, and experiences mental and physical deterioration. Drug-related criminal activity is at a somewhat higher level than in the other groups. The age range of these chronic abusers was between 20 and 28 years. McSherry estimated that the majority (60%) of the population seen in treatment were chronic abusers; the experimental and acute abusers constituted approximately 20% each (McSherry, 1988).

Oetting et al. (1988) have independently defined three types of solvent abusers in the populations that they have studied. These types have been defined through examination of a variety of research data, the majority of which were survey and epidemiological studies. These three main types of inhalant abusers are similar to McSherry's and are referred to as: (a) inhalant-dependent adults, (b) polydrug abusers, and (c) young inhalant abusers. According to Oetting et al. nearly all inhalant abusers in the United States can be categorized into one of these three groups.

Research on inhalant/solvent abuse over a number of years at the Center on Alcoholism, Substance Abuse, and Addictions, at the University of New Mexico, support the three types proposed by McSherry (1988) and Oetting et al. (1988). Furthermore, these three types, when viewed longitudinally, describe inhalant abuse over one's lifespan. The following section provides an attempt to validate the findings of McSherry and Oetting et al. with findings from research on inhalant/solvent abuse conducted by the authors.

### *Experimental, Vicarious Youthful Use*

It was our experience that many children in the western United States experiment infrequently with inhalants and solvents in their preteen and early teen years. In many cases, curiosity about being high on any substance, attention seeking, and peer group influence are the major determining factors. The supply of available solvents is often a factor in experimentation, and solvent use provides a vicarious (substitute) high when intoxicants such as alcohol and other drugs are not available to these youths. This kind of experimentation may be more common in rural areas than in urban centers, although some localized ethnic neighborhoods may have problems as well (De la Rosa et al. 1990; Padilla et al. 1979). Such behaviors tend to be particularly more common in specific peer groups located in particular communities, neighborhoods, ethnic groups, and schools. Oetting and Webb (1992) refer to these environmental influences as causing "hot spots" of high inhalant use. The Canadian literature (Barnes, 1980; Remington & Hoffman, 1984; Smart, 1988) also indicates that these social phenomenon can contribute to a substantially high prevalence in some isolated Native communities, producing lifetime use prevalence rates as high as 60-80% among school-age youths. Barnes's surveys, however, found that only one of six Native villages had a high prevalence (>10%) of either gasoline inhalation or other solvent use among youths (Barnes, 1980).

In describing this pattern, Oetting et al. (1988) indicate that young children are merely experimenting at this stage and, therefore, dependence on the solvents or need for their regular use is not likely. Because this is a child's first introduction to inhalants or solvents, or both, as well as to drugs in general, there may be little long-term commitment to the use of inhalants and solvents at this stage. In most cases, the substances used in these experiences are gasoline, glue, correction fluid, and spray paint, although substances such as other aerosols and cleaning fluids have been used regularly.

### *Inhalant Abuse as Part of Polysubstance Abuse*

As children progress through their midteens (ages 15 through 18, or grades 9 through 12), access to and experience with other drugs of abuse become common; and to a great degree these drugs replace solvents as vehicles of intoxication and euphoria. Many youth who gravitate to peer clusters that emphasize the recreational use of drugs have experience with inhalants and solvents, among a variety of other drugs. The prevalence of inhalant-other drug users, however, is lower than the prevalence of experimental users. In polysubstance abuse, inhalants are part of the sequence of use with alcohol and other drugs. The youths are seeking special intoxicating effects and sensations that they perceive as not forthcoming from other aspects of their lives. In many cases, youths in this category will use almost any type of drug available, and solvents and inhalants become special attractions when other activities or when alcohol or other recreational drugs are not available.

Oetting et al. (1988) characterize these individuals as adolescents who use drugs frequently and for whom drugs play a major part in their lives. They point out that not all polysubstance abusers use inhalants, but some peer clusters of polysubstance abusers do adopt and frequently use inhalants and solvents. Oetting et al. (1988) also indicate that many of these polysubstance abusers have multiple problems. Inhalant abuse is linked more specifically to family problems, social problems, school problems, and psychological problems than is true for experimental users. Furthermore, researchers have indicated that although polydrug abusers who use inhalants rarely exhibit solvent-related medical problems that result in the need for emergency medical care, they are likely headed for future drug-related problems.

### *Adult Chronic Users*

For this category we utilize a title similar to that of Oetting et al. (1988). This group was characterized by both McSherry (1988) and Oetting et al. (1988) as having used inhalants and solvents for many years, and for whom inhalants and solvents have become the drug of choice. Such individuals are quite frequently high on inhalants (generally daily), and they stay chronically intoxicated. Their lives are centered, to a great degree, around inhalant and solvent use, ranging from acquisition to techniques of consumption.

People in this particular category have serious social problems. They are frequently unemployed, even though they are in their twenties or older. They often have medical or psychological problems that bring them in contact with emergency rooms and mental health units. Abuse by inhalant-dependent adults is by far the rarest form of inhalant abuse in the United States, although it is the most likely to be recognized and referred to treatment. We will examine estimates of its prevalence in the following sections.

The adult chronic inhalant abuser is a social isolate in later life, and may also have been a social isolate early in life. Even though these people may have been introduced to inhalants through a peer group, they have little or no peer identification as they get into the chronic phase. Many may have come from disrupted, isolated, multiproblem families, and they generally do not achieve independence for any length of time as adults. In the end stage of use, many wind up in emergency rooms, as inpatients in intensive medical care units (Streicher, Gabow, Moss, Kono, & Kaehug, 1981), in psychiatric hospitals, or in rest-home care, by the time they reach their thirties.

At the end stage of the cycle of adult chronic use, the risk of death is high. Although major medical problems and sudden sniffing death (Anderson, MacNair, & Ransay, 1985; Bass, 1970; Coulehan et al. 1983; Garriott & Petty, 1980) can result in mortality at any age, it is quite likely in chronic users. Garriott and Petty (1980) reviewed the deaths of 34 adults from inhalation of solvents and toxic substances. Contained in this sample were deaths of a number of individuals who were 20 years of age or older from inhalation of spray paint, lighter fluid, propane aerosols, propane, and freon.

Figure 1 provides a graphic illustration of the hypothetical life course (career) of inhalant abuse.

### *Discussion of the Three Types*

Although the literature is complete with descriptions of inhalant abusers in general, an examination indicates that most of the studies that define traits of inhalant abusers are concerned with the second behavioral type discussed, that is, the adolescent polysubstance abuser who uses a substantial amount of inhalants. Most of the survey literature deals with inhalant abusers involved in vicarious use and polysubstance abuse. The more detailed and analytical literature is concerned with polysubstance abuse and adult use. There is very little explicit literature on what separates vicarious use as youngsters from later polysubstance abusers who use inhalants. And there are few studies describing adult chronic abusers.

Beauvais (1992b) noted that with the exception of the National Household Survey in the United States, most evidence for adult use of solvents is anecdotal. Hersey and Miller (1982), in an article that presents an exception to this statement, reviewed drug-regulated emergency room visits for a four-year period. They found six (4%) adult cases of inhalant abuse among the 160 cases of drug-related admissions. These findings, although based on small, non-random samples, lend credence to identifying patterns of solvent abuse among adults. McSherry (1988) also emphasized the need for further study of adults, for they are the majority of the treated population of solvent abusers. Recently, in the popular press of New Mexico, substantial attention has been paid to adult solvent use (Albuquerque Tribune, 1988; Guthrie, 1995).

FIGURE 1. Hypothetical Life Course (Career) of Inhalant Abuse

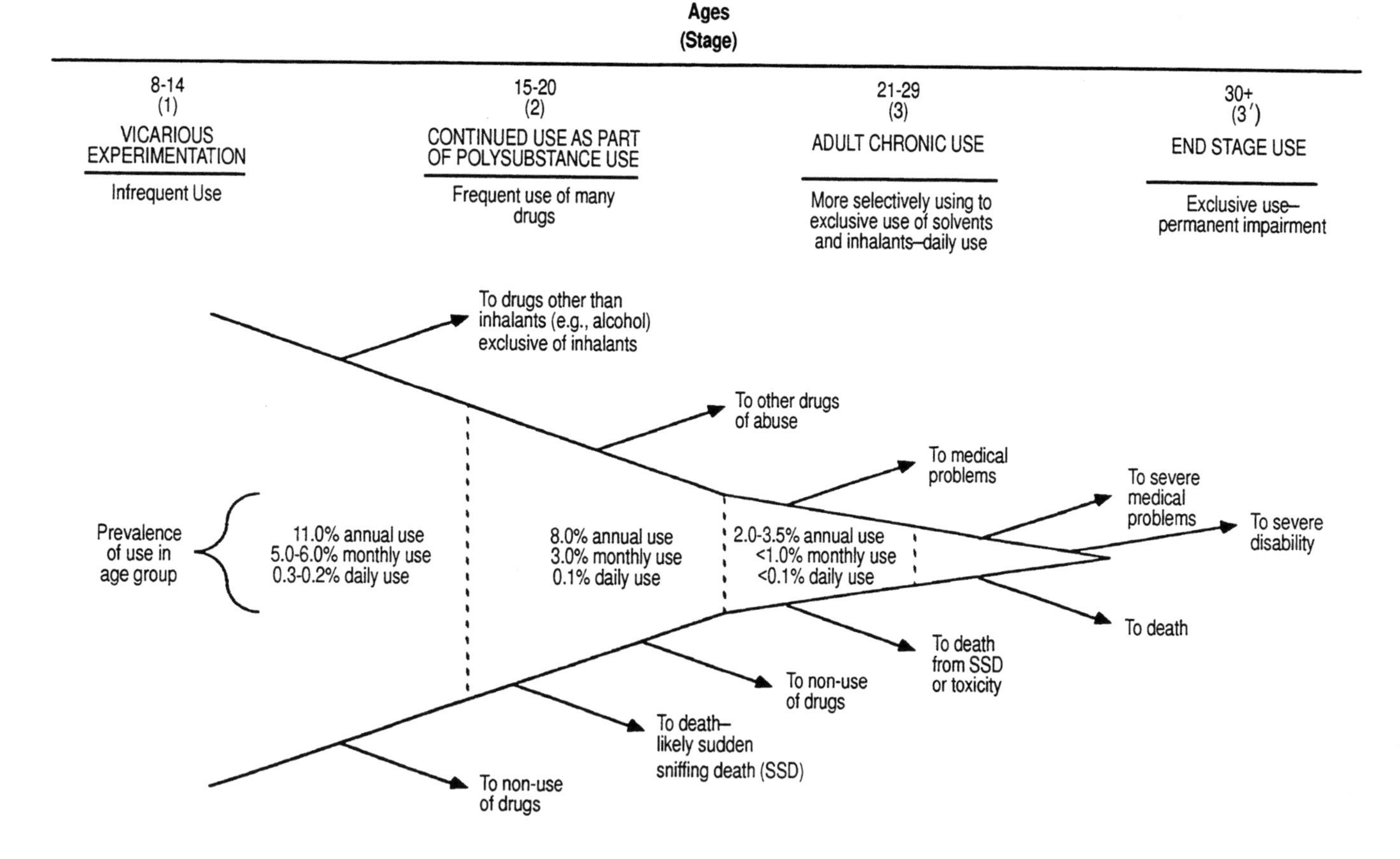

## *DATA SOURCES*

The data that we examine in this section comes from several sources and provides evidence for the three behavioral types of inhalant abuse. First, data from the National Household Survey of the National Institute on Drug Abuse (NIDA, 1994) will be revisited, as well as data from the Monitoring the Futures study carried out by the University of Michigan Institute for Social Research, also for NIDA (Johnston, O'Malley, & Bachman, 1994). Second, other data sources that originate from our research are detailed.

The data presented from New Mexico comes from three sources: two school surveys in rural and small-town areas in various parts of New Mexico, and one random household survey of adult prevalence of drinking and drug use in a wage-work community of less than 10,000 people on a large American Indian reservation in New Mexico.

The high school survey was conducted as part of the Native American Adolescent Injury Prevention Program, which was funded by a federal grant to the New Mexico State Department of Health. It was piloted in 1986 and implemented between 1986 and 1988; it consisted of three waves of data from three different school years. The sample comprised 5,891 students from the 6th through the 12th grades at 12 rural middle and high school sites, and small urban school districts in New Mexico. The rural school districts made up 32.5% of the sample, and the small urban districts 67.4%. The city populations ranged from 2,615 to 19,154. The respondents participated in a survey of drug use and safety practices for the evaluation of an injury prevention program. The age range was from 10 years to 20 years; the mean age was 14.4 years. The participants' mean school grade was 9. Gender participation was almost equal (female students = 49.9%; males = 50.1%). The data presented here are from wave three, collected in the spring of 1988. The schools surveyed were located in western New Mexico, central New Mexico, and southeastern New Mexico. The students represent a tri-ethnic population of 1,270 Anglos (White non-Hispanics), 979 Hispanics, and a large number of students from three American Indian tribes (Navajo, 2,717; Apache, 286; and Pueblo, 510). All of the surveys were designed to monitor alcohol and substance use among the high school students as part of an effort to elicit and describe a number of risk factors for injury.

The second school sample comes from a project entitled Voices of Indian Teens. This multisite project which focused on high school students has been initiated in a number of areas of the western United States. The project has been funded by a grant from the National Institute on Alcohol Abuse and Alcoholism to the National Center for American Indian and

Alaska Native Mental Health Research at the University of Colorado Health Sciences Center. We were co-investigators in this study and the site for data collection and analyses for high schools in New Mexico. The five high schools that participated in the survey were located in rural areas or small towns in western and central New Mexico. All of the high schools had students who were Indian, Hispanic and Anglo, but four of the high schools were populated predominantly by American Indians of Pueblo descent, while the fifth one was predominantly Navajo. The survey instrument was designed to describe and monitor the use of alcohol and drugs and the concurrent mental health status of Indian teens, and to further define the etiology of problems in these topical areas. The data presented are from wave three, carried out in the fall of 1993. The total sample is 900 students.

The adult data originate from a substance use survey of a small town in western New Mexico. This wage-work town is predominantly American Indian, largely from a single tribe, with a population of approximately 6,000. The sample comprised 167 adults, 18 years of age or older, who were interviewed in 1993. The sample was random and was done on a household basis. The various households were located on a map and chosen by a random number system. Over two thirds of the sample was female. Interviews were conducted during work hours.

## *EVIDENCE FOR THE THREE TYPES OF INHALANT ABUSERS*

Before turning to the New Mexico data, a brief review of the national data contained in Tables 1 and 2 is important. Observation of these data shows that inhalant abuse diminishes with age in the general population; it is much more common in the early high school years, and tapers off to become quite rare in adults. The pattern shown by the national household data is consistent with the notion that substance abuse becomes less frequent with age, a progression corresponding to the three different types of inhalant users. The occasional inhalant users raise the "used in the past year" and "past month" estimates in the younger age categories, but as these occasional users quit or move to other drugs, then the lower percentages of "used in past year" and "past month" represent only the people progressing toward the chronic adult pattern.

The Monitoring the Futures study data (Johnston et al. 1994) found in Table 2 are probably more useful for our purposes. The highest prevalence rates in all of the various use categories (lifetime, annual, 30-day, and daily) are found in the eighth grade; in the older grade and age groups the prevalence of use on all of the prevalence measures is reduced substantially. The occasional use gradually gives way to polysubstance abuse, which continues in 10th and 12th grades and in college. By the time young adulthood is reached, however, regular use by a substantial proportion of the population is quite rare. For example, 30-day use in 1993 decreased from 5.4% to 0.7% from eighth grade to the young adult years. It is even more important to note that very few people become involved in the chronic adult use pattern, and that the daily use in 1993 never exceeded 0.3% in any age group. Use by the young adult group was found to be less than 0.05%. One advantage of these data is that the young adult survey was population based rather than school based. On the other hand, it is likely that adult use is underreported by this survey, given the social characteristics (e.g., isolation and marginality) of chronic adult users.

The national data do not tell us a great deal about the three types of inhalant abuse. However, the data trends are such that they support, or at least do not contradict, the existence of the three patterns. It is likely that the three patterns of behavior exist within the data, but to elicit the actual types from national data sets is not possible without new approaches to the questions asked, and without major multivariate, longitudinal research designs.

### *New Mexico Middle and High School Survey*

Table 3 provides a comparison of the demographic characteristics of tri-ethnic middle school and high school inhalant users with the non-inhalant users in the sample.

The Indian and Hispanic inhalant users are younger than the overall samples within their own ethnic groups; however, the Anglo inhalant users are older than the few number of Anglos. Male-female use is about equal in both Hispanic and Indian users and non-users, but Anglo inhalant users are predominantly male. Use of inhalants peaks in 8th grade for the Indian inhalant users, but some use continues through the 12th grade. Among the Hispanics, the peak years are eighth and ninth grades, with a greater degree of falloff in the later years. Finally, the inhalant users among the Anglos are quite evenly distributed from 8th through 12th grades. In all three ethnic groups the inhalant abusers have lower aspirations for future schooling than do others in their own ethnic groups. Finally, the respondents' reports of grades earned show no clear pattern for either Hispanics or Anglos. In these two groups, the grades earned do not seem to be substantially different from the grades of others within their ethnic group. However, among the Indian respondents, the grades earned by inhalant users are slightly lower than the grades of others in this group.

Data from the same sample presented in Table 4 provide other insights

TABLE 3. Demographic Characteristics for Inhalant Users (in the Past Week) and Non-Users in a Tri-Ethnic Middle and High School Sample, 1988

| | **American Indian Total** (N = 3527) | **American Indian Inhalant** (N = 114) | **Hispanic Total** (N = 979) | **Hispanic Inhalant** (N = 25) | **Anglo Total** (N = 1270) | **Anglo Inhalant** (N = 23) |
|---|---|---|---|---|---|---|
| **Age (years)** | | | | | | |
| Mean | 14.4 | 13.8 | 14.1 | 13.8 | 14.2 | 15.2 |
| SD | 2.0 | 1.7 | 1.9 | 1.5 | 1.9 | 1.6 |
| Range | 10-20 | 10-18 | 10-20 | 11-17 | 10-19 | 13-18 |
| **Gender (%)** | | | | | | |
| Male | 50.0 | 59.0 | 51.0 | 56.0 | 52.0 | 87.0 |
| Female | 50.0 | 41.0 | 49.0 | 44.0 | 48.0 | 13.0 |
| **Grade (%)** | | | | | | |
| 6th | 11.0 | 19.0 | 13.0 | 8.0 | 13.0 | 0.0 |
| 7th | 15.0 | 18.0 | 15.0 | 16.0 | 14.0 | 0.0 |
| 8th | 16.0 | 22.0 | 19.0 | 28.0 | 14.0 | 22.0 |
| 9th | 17.0 | 15.0 | 19.0 | 28.0 | 15.0 | 22.0 |
| 10th | 16.0 | 10.0 | 15.0 | 12.0 | 15.0 | 13.0 |
| 11th | 14.0 | 10.0 | 11.0 | 4.0 | 15.0 | 22.0 |
| 12th | 12.0 | 6.0 | 9.0 | 4.0 | 14.0 | 22.0 |
| **How far do you plan to go in school (%)** | | | | | | |
| some high school | 2.0 | 4.0 | 1.0 | 12.0 | 1.0 | 4.0 |
| high school graduate | 35.0 | 46.0 | 34.0 | 44.0 | 16.0 | 26.0 |
| vocational/technical | 7.0 | 4.0 | 7.0 | 0.0 | 6.0 | 13.0 |
| some college | 15.0 | 11.0 | 14.0 | 12.0 | 10.0 | 17.0 |
| college graduate | 26.0 | 21.0 | 28.0 | 28.0 | 42.0 | 22.0 |
| some graduate school | 2.0 | 1.0 | 1.0 | 0.0 | 1.0 | 0.0 |
| graduate degree | 14.0 | 11.0 | 15.0 | 4.0 | 25.0 | 13.0 |
| **Grades usually earned in school (%)** | | | | | | |
| mostly A's | 9.0 | 8.0 | 4.0 | 9.0 | 22.0 | 13.0 |
| mostly B's | 41.0 | 33.0 | 36.0 | 45.0 | 49.0 | 26.0 |
| mostly C's | 42.0 | 41.0 | 44.0 | 40.0 | 25.0 | 52.0 |
| mostly D's | 7.0 | 15.0 | 16.0 | 4.0 | 2.0 | 9.0 |
| mostly F's | 1.0 | 2.0 | 0.0 | 0.0 | 1.0 | 0.0 |

into drug use and family problems. In all three ethnic groups the inhalant users are 40-75% more likely than non-users to have been drunk in their lifetimes. Furthermore, the inhalant users are 2 to 3 times more likely to have been drunk in the last week. The inhalant users are also 3 to 8 times more likely to have used marijuana in the past week, with the Anglos showing the greatest marijuana use.

Responses to the final question on drug use in Table 4, regarding inhalant use, indicate that 3.2% of the overall sample of American Indians used inhalants in the last week, whereas 2.6% of the Hispanics and 1.8% of the Anglos did so.

Table 4 also presents variables on family problems. Responses indicate that the Indian and Anglo inhalant users are more likely to have parents who drink and drive with their children than do members of the Hispanic sample. The data on unemployment vary in the samples, but overall they are equal. Approximately 14% of all the parents or grandparents of the Indian, Anglo, and Hispanic samples are reported to have problems with alcohol, but the disparity between the inhalant users and the regular sample is much higher among the Anglos than it is within the other groups. Twenty-six percent of the Anglo inhalant-users' parents are reported as having problems with alcohol. Separation and divorce are also higher in the Anglo inhalant-user sample than in the regular sample, whereas these behaviors are not substantially higher among the inhalant users of the other two ethnic groups. Finally, using foster placement as an indicator of family disruption, we find that problems seem to exist in all three ethnic groups. In each of the three groups of students, the inhalant users are 2 times more likely to have had siblings placed in a foster home than the overall samples of their particular ethnic groups.

### *New Mexico High School Survey, 1993*

From a more recent survey of the demographic characteristics of southwestern American Indian and non-Indian high school samples, we gain insights similar to those derived from Table 4. Table 5 provides a comparison of inhalant users and frequent inhalant users with the total sample of these particular ethnic groups. Among American Indians the inhalant users are younger than the other Indians in the sample, more likely to be female, and found more commonly in the lower grades. The frequent users, in particular, are concentrated in the lower grades, primarily ninth and tenth. As far as their plans for the future are concerned, they do not differ greatly from the overall Indian sample, and the grades that they report earning are also about the same as the total sample.

The non-Indian sample presents a slightly contrasting, but not entirely

TABLE 4. Drug Use and Family Problems for Inhalant Users (in the Past Week) and Non-Users in a Tri-Ethnic Middle and High School Sample, 1988

| | American Indian Total (N = 3527) | American Indian Inhalant (N = 114) | Hispanic Total (N = 979) | Hispanic Inhalant (N = 25) | Anglo Total (N = 1270) | Anglo Inhalant (N = 23) |
|---|---|---|---|---|---|---|
| **Other Drug Use (% yes)** | | | | | | |
| Have you ever been drunk? | 49.0 | 69.0 | 56.0 | 80.0 | 52.0 | 91.0 |
| Were you drunk last week? | 14.0 | 33.0 | 15.0 | 40.0 | 14.0 | 48.0 |
| Ever had more than 2 beers to get high? | 39.0 | 61.0 | 46.0 | 72.0 | 43.0 | 83.0 |
| Were you high on marijuana last week? | 11.0 | 35.0 | 9.0 | 28.0 | 6.0 | 57.0 |
| Were you high on inhalants last week? | 3.2 | 100.0 | 2.6 | 100.0 | 1.8 | 100.0 |
| **Family Problems (% yes)** | | | | | | |
| Ever ride with a parent drinking alcohol? | 24.0 | 41.0 | 30.0 | 28.0 | 28.0 | 48.0 |
| Was either parent unemployed in the past year? | 35.0 | 37.0 | 30.0 | 12.0 | 31.0 | 44.0 |
| Have your parents or grandparents ever had a problem with alcohol? | 13.0 | 18.0 | 14.0 | 12.0 | 14.0 | 26.0 |
| Are your parents separated or divorced? | 30.0 | 35.0 | 34.0 | 28.0 | 39.0 | 52.0 |
| Have you or your siblings ever been placed in a foster home? | 7.0 | 18.0 | 2.0 | 4.0 | 2.0 | 4.0 |

different, picture. The non-Indian inhalant users are not really different in age from the overall non-Indian sample. Furthermore, their grade pattern is not substantially different. Although there are more females in the inhalant-user category, the sex ratio is closer to equal than in the Indian sample. In the self-assessment items, however, the non-Indian sample is quite different. The non-Indian inhalant users report substantially lower educational aspirations as well as lower grade performance. As others have suggested regarding all drugs of abuse (Winfree & Griffiths, 1983; Winfree, Griffiths, & Sellers, 1989), it is possible that the non-Indian drug users, in this case inhalant users, are much more likely to assume a deviant identity than are American Indian users.

In Table 6, other drug use is compared with inhalant-using and non-inhalant-using Indians and non-Indians. Polysubstance abuse of the inhalers is highly supported by these data.

In the Indian sample, both in the "ever used" and "use in last month" sample, the inhalant users reported substantially higher use of every category of drug. For marijuana, cocaine, solvents, barbiturates, and others, the Indian inhalant users generally reported 2 times or greater use for both the "ever used" and "frequent use" categories. For example, cocaine use is 2.5 to 7 times higher in the "ever used" category for the inhalant users. Furthermore, Indian "frequent users" have higher rates of use in every category of drugs than do the overall samples and the other inhalant users. Finally, the age of first use does not greatly distinguish Indian inhalant users from non-inhalant users; the averages are only very slightly lower for the inhalant users.

In Table 6, the data for the non-Indians exhibit similar patterns to data for the Indians, except that the magnitude of polysubstance use is even greater among the non-Indian inhalant users. For example, the use of barbiturates is 4 to 5 times higher among the inhalant-using non-Indians. Use in the last month among the non-Indians further portrays the non-Indian inhalant users as frequent and heavy polysubstance abusers. Finally, the age of first use among the non-Indian sample of inhalant users is lower for virtually all drugs than is age of first use in the overall non-Indian sample. But it is not dramatically different for any drug except "speed" among the frequent inhalant users.

In Table 7, a question which ostensibly separates polysubstance-abusing adolescents from those who will go on to become adult chronic users, is presented. Students were asked, "If alcohol were available, would you use any of the above drugs?" The expectation is that inhalant users who go on to become chronic inhalant/solvent users would be less likely to give up inhalants and solvents once alcohol becomes readily available. The data

TABLE 5. Selected Demographic Characteristics for Southwestern American Indian and Non-Indian School Samples, Inhalant Users, and Frequent Inhalant Users, 1993

| | American Indian Total (N = 694) | American Indian Inhalant (N = 127) | American Indian Frequent (N = 94) | Non-American Indian All (N = 199) | Non-American Indian Inhalant (N = 24) | Non-American Indian Frequent (N = 16) |
|---|---|---|---|---|---|---|
| **Age (years)** | | | | | | |
| Mean | 16.0 | 15.4 | 15.4 | 15.7 | 15.6 | 16.0 |
| SD | 1.4 | 1.3 | 1.3 | 1.3 | 1.4 | 1.2 |
| Range | 13-21 | 13-19 | 13-19 | 13-18 | 13-18 | 14-18 |
| **Gender (%)** | | | | | | |
| Male | 51.0 | 37.0 | 42.0 | 48.0 | 48.0 | 41.0 |
| Female | 49.0 | 63.0 | 59.0 | 52.0 | 52.0 | 59.0 |
| **Grade (%)** | | | | | | |
| 9th | 26.0 | 38.0 | 42.0 | 32.0 | 38.0 | 25.0 |
| 10th | 24.0 | 28.0 | 29.0 | 26.0 | 25.0 | 31.0 |
| 11th | 25.0 | 22.0 | 18.0 | 20.0 | 21.0 | 25.0 |
| 12th | 23.0 | 9.0 | 10.0 | 18.0 | 13.0 | 13.0 |
| **How far do you plan to go in school? (%)** | | | | | | |
| some high school | 1.0 | 1.0 | 1.0 | 3.0 | 13.0 | 19.0 |
| high school graduate | 32.0 | 37.0 | 40.0 | 23.0 | 38.0 | 44.0 |
| vocational/technical | 13.0 | 10.0 | 10.0 | 15.0 | 13.0 | 13.0 |
| military training | 10.0 | 10.0 | 11.0 | 9.0 | 4.0 | 0.0 |
| four-year college | 26.0 | 24.0 | 21.0 | 31.0 | 17.0 | 6.0 |
| graduate degree | 17.0 | 17.0 | 17.0 | 39.0 | 17.0 | 19.0 |
| **Grades usually earned in school (%)** | | | | | | |
| mostly A's | 11.0 | 12.0 | 12.0 | 22.0 | 8.0 | 6.0 |
| mostly B's | 45.0 | 39.0 | 38.0 | 40.0 | 33.0 | 38.0 |
| mostly C's | 38.0 | 38.0 | 38.0 | 29.0 | 38.0 | 31.0 |
| mostly D's and lower | 6.0 | 10.0 | 12.0 | 7.0 | 21.0 | 25.0 |

TABLE 6. Rates of Drug Use, Frequency of Drug Use, and Mean Age First Used Drugs for Southwestern American Indian and Non-Indian School Samples, Inhalant Users, Frequent Inhalant Users, 1993

| | American Indian Total (N = 694) | American Indian Inhalant (N = 127) | American Indian Frequent (N = 94) | Non-American Indian All (N = 199) | Non-American Indian Inhalant (N = 24) | Non-American Indian Frequent (N = 16) |
|---|---|---|---|---|---|---|
| **Have you ever used (% yes)** | | | | | | |
| Marijuana | 61.0 | 85.0 | 84.0 | 42.0 | 75.0 | 81.0 |
| Crack or cocaine | 10.0 | 25.0 | 73.0 | 9.0 | 29.0 | 31.0 |
| Inhalants | 18.0 | 100.0 | 100.0 | 12.0 | 100.0 | 100.0 |
| Solvents (drinking) | 7.0 | 23.0 | 25.0 | 8.0 | 38.0 | 44.0 |
| Amphetamines or speed | 17.0 | 43.0 | 44.0 | 13.0 | 29.0 | 31.0 |
| Barbiturates | 7.0 | 22.0 | 23.0 | 8.0 | 38.0 | 31.0 |
| Other drugs (heroin, LSD, PCP, Ecstasy, or MDA) | 11.0 | 25.0 | 23.0 | 12.0 | 42.0 | 50.0 |
| **How many times in the last month did you use (mean # of times)** | | | | | | |
| Marijuana | 8.1 | 10.1 | 11.1 | 8.4 | 14.9 | 19.5 |
| Crack or cocaine | 4.6 | 5.9 | 7.2 | 5.8 | 12.5 | 12.0 |
| Inhalants | 3.6 | 4.2 | 5.4 | 4.7 | 6.5 | 8.9 |
| Solvents (drinking) | 2.8 | 4.1 | 4.9 | 5.4 | 11.6 | 14.2 |
| Amphetamines or speed | 3.8 | 4.4 | 5.5 | 4.3 | 13.0 | 17.8 |
| Barbiturates | 3.7 | 5.5 | 6.7 | 5.1 | 10.9 | 12.2 |
| Other drugs (heroin, LSD, PCP, Ecstasy, or MDA) | 3.8 | 5.5 | 6.8 | 4.6 | 9.1 | 11.3 |
| **How old were you when you first tried (mean age)** | | | | | | |
| Marijuana | 12.9 | 12.5 | 12.5 | 13.3 | 12.5 | 13.2 |
| Crack or cocaine | 15.0 | 14.3 | 14.3 | 13.1 | 13.7 | 12.8 |
| Inhalants | 13.1 | 13.1 | 13.2 | 12.6 | 12.5 | 12.9 |
| Solvents (drinking) | 13.2 | 13.1 | 13.1 | 12.8 | 11.9 | 11.9 |
| Amphetamines or speed | 14.1 | 14.0 | 14.2 | 13.2 | 11.0 | 10.3 |
| Barbiturates | 14.2 | 13.4 | 13.8 | 13.3 | 12.5 | 12.8 |
| Other drugs (heroin, LSD, PCP, Ecstasy, or MDA) | 14.3 | 13.9 | 14.4 | 13.6 | 13.8 | 13.6 |

TABLE 7. Alcohol, Problems Associated with Alcohol Use, and Peer Use of Alcohol for Southwestern American Indian and Non-Indian School Samples, Inhalant Users, and Frequent Inhalant Users, 1993

| | American Indian Total (N = 694) | American Indian Inhalant (N = 127) | American Indian Frequent (N = 94) | Non-American Indian All (N = 199) | Non-American Indian Inhalant (N = 24) | Non-American Indian Frequent (N = 16) |
|---|---|---|---|---|---|---|
| **If alcohol were available, would you use any of the drugs listed above (% yes)** | 21.0 | 44.0 | 47.0 | 22.0 | 50.0 | 63.0 |
| **Have you ever had a drink of alcohol? (% yes)** | 77.0 | 91.0 | 90.0 | 77.0 | 100.0 | 100.0 |
| **How old were you when you first started drinking? (mean yrs.)** | 13.1 | 12.2 | 12.1 | 12.7 | 12.4 | 12.8 |
| **How old were you when you first got drunk? (mean yrs.)** | 13.3 | 12.7 | 12.6 | 11.9 | 12.2 | 13.3 |
| **How many days did you drink last month? (mean # days)** | 4.0 | 4.9 | 4.9 | 5.0 | 8.3 | 9.2 |
| **How many times did you get drunk in the last month? (mean # times)** | 2.8 | 4.0 | 4.5 | 2.0 | 3.5 | 4.8 |
| **In the past month how often did you . . . (% responded often or almost always)** | | | | | | |
| drink more than you thought you would? | 17.0 | 36.0 | 37.0 | 12.0 | 32.0 | 36.0 |
| have a hangover? | 11.0 | 23.0 | 25.0 | 8.0 | 20.0 | 18.0 |
| go to class drunk or hungover? | 5.0 | 13.0 | 16.0 | 6.0 | 20.0 | 18.0 |
| your grades go down due to drinking? | 5.0 | 11.0 | 11.0 | 8.0 | 20.0 | 24.0 |
| **How often do your friends . . . (% responded often or almost always)** | | | | | | |
| ask you to get drunk? | 19.0 | 37.0 | 37.0 | 20.0 | 60.0 | 65.0 |
| try to stop you from getting drunk? | 10.0 | 12.0 | 14.0 | 8.0 | 24.0 | 29.0 |
| **How often do you try to stop friends from getting drunk? (% responded often or almost always)** | 15.0 | 13.0 | 15.0 | 14.0 | 16.0 | 24.0 |
| **How many of your friends get drunk? (% responded some or a lot)** | 54.0 | 70.0 | 71.0 | 51.0 | 84.0 | 88.0 |

support this notion, in that among the Indian sample, twice as many of the inhalant users and frequent users would not give up other drugs for alcohol. Similarly, among the non-Indian sample, 2 to 3 times as many non-Indian inhalant users would not give up other drugs for alcohol.

Table 7 also presents other indicators of drinking and heavy drinking. Generally the Indian inhalant abusers are about the same as the non-inhalant users on measures such as first experience with drinking, being drunk, and their experience in drinking last month. However, when asked about ever drinking, their rates are slightly higher; and when being drunk in the last month is measured, the rates are substantially higher. It is interesting to note, however, that 9-10% of the Indian inhalant users do not drink at all. These may be individuals who will go on to be isolated, chronic inhalant users.

Examining these same variables for non-Indians, we find that inhalant user's age at first drinking is about the same as the entire sample of non-Indians, and their age at first being drunk is older. However, on the three other variables ("ever drink," "drunk last month," and "drink last month") they are all substantially higher. Furthermore, all of the non-Indian inhalant users have experienced drinking alcohol.

Table 7 also indicates that the Indian inhalant users have much heavier alcohol-use indicators than does the total Indian sample. The indicators of heavy use are generally 2 times higher than in the total Indian sample. Among non-Indians, the same indicators of heavy use are 3 times higher for the non-Indian inhalant users than for the entire non-Indian sample.

The next variables presented in Table 7 concern the social pressures for drinking and nondrinking among the respondents' friends. Indian inhalant users report that their friends are more likely to encourage or support drinking than is the entire Indian sample. Non-Indian inhalant users are even more likely to report strong peer support for drinking. However, they are also more likely to report that their friends would try to stop them from drinking. Therefore, it appears that the non-Indian inhalant users have friends who support nondrinking as well as drinking, whereas the Indian inhalant users have friends who are more likely to support drinking and also not to stop them from drinking. Finally, the last item shows that Indian inhalant users estimate that a lower proportion of their friends get drunk, whereas the non-Indian users report a perception that a very high percentage of their friends get drunk.

Table 8 contains responses on suicidal ideation which indicate that both Indian and non-Indian inhalant users and frequent users are consistently more suicidal at all levels–from having thought about killing themselves to planning an attempt. In the second part of Table 8 are data on social support items. In general there is little difference in respondents' perceptions of social support, but one exception is family communication. Fewer of the inhalant abusers, particularly the non-Indians, report that they can talk with their families when compared to those who do not use inhalants.

In the final variables in Table 8, community and family problems are highlighted. For both the Indian and non-Indian samples, the inhalant abusers within each ethnic group are more likely to drink with older siblings. The Indian inhalant abusers are 2 times more likely than the total Indian sample, and the frequent inhalant users among the non-Indians are 3 times more likely than the total non-Indian sample to drink with siblings. Both Indian and non-Indian inhalant users perceive the community as supporting adolescent drinking. Again, the frequent users among the non-Indians are most likely to agree with this statement. Finally, the inhalant abusers among both ethnic groups are more likely than the total samples to have an important adult with alcohol problems. Responses indicate that 12-17% more of the inhalant users are more likely to have this experience. Finally, there is no difference in the unemployment experience of the parents of inhalant users and non-users in the two ethnic groups.

***Adult Substance Use Survey***

Turning now to the American Indian adult community sample (Table 9), we will review evidence for adult inhalant patterns. There is no major difference in the age of inhalant users and the age of the overall sample, although inhalant users are slightly younger. By the time people reach adult status, it appears that inhalant users are exclusively male. In educational and occupational variables, there is no major difference between the inhalant users and the overall adult sample. On marital status, however, the current inhalant users are much less likely to be married–50% are single (versus 29% of the overall sample) or divorced (25% versus 10%).

Table 10 highlights drugs used by adults. Inhalant users are more likely to use all common drugs (both ever used and current user), with current users showing higher rates on all categories with the exception of prescription drugs. Current inhalant users are much more likely to use alcohol, cigarettes, and marijuana. Although the results are not contained in Table 10, individuals were asked which drugs were hard to cut down on or to give up. Of the current inhalant users, 25% reported difficulty in cutting down on alcohol and inhalants, but they did not report difficulty in cutting down with any others. Of the ever users, 14% said that they had trouble cutting down on inhalant use.

Other indicators of alcohol and drug use behavior are also presented in Table 10. Age at first use of alcohol and age of regular use of alcohol are

TABLE 8. Suicidal Ideation, Social Support, and Community Family Problems for American Indian and Non-Indian School Samples, Inhalant Users, and Frequent Inhalant Users, 1993

| | **American Indian Total** (N = 694) | **American Indian Inhalant** (N = 127) | **American Indian Frequent** (N = 94) | **Non-American Indian All** (N = 199) | **Non-American Indian Inhalant** (N = 24) | **Non-American Indian Frequent** (N = 16) |
|---|---|---|---|---|---|---|
| **Suicidal ideation (% reported in past month or more often)** | | | | | | |
| **I have thought about . . .** | | | | | | |
| killing myself | 11 | 24 | 26 | 13 | 28 | 35 |
| how to kill myself | 11 | 24 | 26 | 11 | 20 | 24 |
| when I would kill myself | 10 | 22 | 22 | 10 | 16 | 18 |
| a suicide note | 7 | 18 | 17 | 9 | 12 | 12 |
| writing a will | 5 | 12 | 14 | 8 | 12 | 12 |
| telling people my suicide plan | 6 | 13 | 14 | 6 | 8 | 12 |
| **Social Support (% agree with this statement)** | | | | | | |
| There is a special person around when I am in need. | 65 | 68 | 70 | 70 | 60 | 65 |
| I have a special person who is a real source of comfort to me. | 68 | 69 | 66 | 71 | 60 | 60 |
| My family really tries to help me. | 65 | 60 | 59 | 66 | 52 | 63 |
| I can talk about my problems with my family. | 43 | 32 | 30 | 50 | 24 | 29 |
| I have friends with whom I can share my joys and sorrows. | 65 | 69 | 65 | 69 | 60 | 56 |
| I can talk about my problems with my friends. | 61 | 65 | 63 | 63 | 64 | 65 |
| **Community/Family Problems (% responded yes)** | | | | | | |
| drink with older siblings | 11 | 19 | 22 | 8 | 13 | 31 |
| community supports adolescent drinking | 7 | 13 | 14 | 13 | 17 | 25 |
| important adult has an alcohol problem | 39 | 56 | 52 | 26 | 38 | 38 |
| parent unemployed | 13 | 16 | 16 | 9 | 8 | 13 |
| parents divorced | 30 | 35 | 35 | 31 | 33 | 31 |

TABLE 9. Demographic Characteristics of an Adult Indian Community Sample Compared to Inhalant Users (Ever Used and Current Users), 1993

| Characteristic | Total Sample (N = 169) | Inhalant Users (N = 6) Ever Used | Inhalant Users (N = 4) |
|---|---|---|---|
| **Age (Years)** | | | |
| Mean | 31.5 | 28.1 | 30.1 |
| SD | 10.5 | 7.5 | 9.0 |
| Range | 16-73 | 18-36 | 22-41 |
| **Gender (%)** | | | |
| Male | 35.0 | 86.0 | 100.0 |
| Female | 60.0 | 14.0 | – |
| **Education (%)** | | | |
| less than high school | 35.0 | 14.0 | – |
| high school graduate | 31.0 | 57.0 | 75.0 |
| GED | 10.0 | – | – |
| Vocational/technical cert. | 8.0 | 28.0 | 25.0 |
| associate degree | 7.0 | – | – |
| college degree | 6.0 | – | – |
| graduate degree | 4.0 | – | – |
| **Occupation (%)** | | | |
| professional | 14.0 | – | – |
| sales | 2.0 | – | 25.0 |
| clerical | 4.0 | – | – |
| blue collar | 16.0 | 43.0 | 25.0 |
| farm worker | 1.0 | – | – |
| student | 5.0 | – | – |
| homemaker | 13.0 | 14.0 | – |
| unemployed | 12.0 | 14.0 | – |
| child care | 8.0 | – | – |
| retired | 1.0 | – | – |
| **Marital Status (%)** | | | |
| single | 29.0 | 29.0 | 50.0 |
| married | 46.0 | 43.0 | 25.0 |
| single with partner | 9.0 | – | – |
| separated | 5.0 | – | – |
| divorced | 10.0 | 29.0 | 25.0 |
| widowed | 1.0 | – | – |

substantially younger for the current adult inhalers. Furthermore, the mean number of times that they drank alcohol in the past month, and the mean number of days drunk in the past year, are higher among the current inhalant abusers. The inhalant users thus show a much higher frequency when compared to non-inhalant users on alcohol indicators. Also, the table reveals that binge-drinking behavior is high among members of the inhal-

TABLE 10. Alcohol and Other Drug Use for an Indian Adult Sample and Inhalant Users (Ever Used and Current Users)

| Drug/Alcohol Behavior | Total Sample (N = 169) | Inhalant Users (N = 6) Ever Used | Inhalant Users (N = 4) Current Users |
|---|---|---|---|
| **Check the drugs you use . . . (% checked)** | | | |
| alcohol | 42.0 | 86.0 | 100.0 |
| cigarettes | 27.0 | 57.0 | 75.0 |
| inhalants | 3.4 | 29.0 | 100.0 |
| marijuana | 12.0 | 43.0 | 50.0 |
| cocaine | 2.4 | – | 25.0 |
| prescription drugs | 5.0 | 14.0 | – |
| heroin | 1.2 | – | 25.0 |
| angel dust | 1.2 | – | 25.0 |
| other | 4.0 | 14.0 | 25.0 |
| **Alcohol Use/Behavior** | | | |
| How old were you when you first began drinking alcohol? (mean years) | 14.0 | 10.7 | 10.3 |
| How old were you when you began to drink once a month? (mean years) | 13.8 | 14.3 | 13.0 |
| How often did you drink during the last month? (mean # of times) | 1.3 | 4.7 | 10.3 |
| How many times have you been drunk in the past year? (mean # of times) | 8.6 | 2.7 | 2.5 |
| Have you ever missed work or school because of a hangover? (% yes) | 18.0 | 86.0 | 2.5 |
| Have you ever been drunk at work or school? (% yes) | 10.0 | 71.0 | 100.0 |
| Have you ever lost a job because of drinking? (% yes) | 12.0 | 86.0 | 75.0 |
| Has your spouse ever told you to cut down on your drinking? (% yes) | 25.0 | 71.0 | 75.0 |

ant using Indian group. Current inhalant users have binged to the point of drunkenness 25 times in the past year.

Current inhalant users are also more likely to have missed work because of a hangover or being drunk at work. They are also more likely to have lost a job because of drinking, and to have been told more frequently by their spouses to cut down on drinking. These findings indicate the presence of heavy drinking among most current inhalant users.

Contained within this sample of 167 adults was one 22-year-old, unemployed male who reported that he used inhalants more frequently than any of the other respondents. He repeatedly used inhalants on Sundays when alcohol was not readily available. Whether this individual represents a current or future adult chronic inhalant and solvent user is not known. However, it is interesting that four (2.4%) of the adults in this overall sample used inhalants in the past year.

## *DISCUSSION*

Having used three samples of youths and adults to describe patterns of inhalant abuse, an analysis of the fit of the data to the three described inhalant abuser types is warranted. Certainly many of the characteristics have been found to be consistent with the descriptions of the three types.

In the tri-ethnic study of middle school and high school students, the early onset of inhalant use is obvious in all of the ethnic groups, particularly among the American Indians and Hispanics. In this sample, an equal number of the males and females, except for Anglos, have used inhalants in the past week. The peak use for all groups is in eighth or ninth grades, and a tapering off of use in the later grades is common. There is major evidence for a pattern of polysubstance abuse among certain groups of adolescents in this population. Therefore, the evidence appears to support the notion of experimental, vicarious use of inhalants in the youngest years, with a movement toward polysubstance abuse in the later grades.

In the recent high school sample covered, heavy polysubstance abuse is very evident among the inhalers. Female inhalers are more prevalent than the male inhalers in this high school sample. This is consistent with some trends documented in other studies in recent years (Beauvais, 1992a). The data, however, raise the suspicion that many, if not most, of the heaviest inhalant users are males. Finally, this data set provides some support for the hypothesis that there is a gradual process whereby many youths give up inhalants for other drugs or behaviors. The inhalant users who continue to use in the later teen years may well be a more troubled group of individuals. For example, we saw in this sample (documented in Table 8) how suicidal ideation is higher, as are a few other measures of psychological disturbance, in the current inhalant users. Furthermore, it is clear that the current inhalant users perceive less social support from their families, a perception that has been reported in other studies. Current inhalant users may well be isolated from their families in their communication quality and patterns. They are, however, perceiving no more dramatic difference in the social support that they receive from their peers than are the other high school youths. Finally, divorce and unemployment do not seem to differentiate the families of the inhalant users from the rest of the sample. Alcohol use with siblings, however, does indicate that more experimentation and learning occur with the support of siblings.

The adult sample is interesting in that heavy polysubstance abuse, including some use of inhalants and solvents, continues into the adult years. Approximately 2% of the adults in our community sample continue to use inhalants and solvents, although most use is very infrequent. Virtually all of these users are males, and most are single or divorced. All of these individuals started use of alcohol and other drugs at an earlier age than the other adults in the sample, most of the inhalant users continue to use other drugs and intoxicants much more frequently than others in the sample as well.

Finally, the social indicators of inhalant users are generally poorer than are those of the rest of the sample. One individual in the sample seemed to fit the pattern of progression toward becoming a chronic adult inhalant user. This person might be classified as a heavy user in the adult stage, yet he is a polysubstance user and does not fit the described pattern of arriving at exclusive solvent use, at least by age 22. In addition, he does not fit the pattern of sustained chronic use.

Overall there was support for the three patterns of inhalant use hypothesized and described by others (McSherry, 1988; Oetting et al., 1988). The data do generally fit the three types of users and the life course that differentiates these three types. However, it is not possible to track inhalant use over the life course with sectional survey data such as we have presented. What we have presented are snapshots from different samples which may serve as pieces of evidence for something that is occurring longitudinally. Further, many forces affect the data which might distort the longitudinal perspective. For example, dropping out of school influences school data (where inhalant-abuser data are lost); but the adult sample is less likely to be affected by social isolation from events such as dropout.

In conclusion, it seems that the existence of the three patterns is modestly supported in these explorations. The data also support what have been our informed observations over the years in some small western communities, and the conclusions of other researchers as well. These three types of use–*experimental use, polysubstance use in adolescence,* and *adult chronic patterns*–do seem to exist. The future, however, holds a number of challenges in further defining these types, the risks associated with each, and the variables linked most closely to individuals who participate in these behaviors.

## FUTURE RESEARCH

Although future research needs concerning inhalant abuse have been detailed elsewhere (Oetting & Webb, 1992; Sharp, Oetting, & Spence, 1992), at least five subtopics involving inhalant/solvent use have been identified that we believe need further exploration: (a) a case control study; (b) an ethnography and social-psychological profile of chronic and adult abusers, including end stage solvent and inhalant abusers; (c) surveillance and documentation of inhalant abuse-related deaths; (d) fetal solvent syndrome; and (e) a longitudinal, repeated measures design. Descriptions of these topics are offered in the following sections.

### Case Control Study

This study would involve comparing a number of chronic and episodic adolescent and young adult inhalant abusers of similar ethnic and community backgrounds with matched controls on a variety of social, psychological, and demographic variables. Controls would be picked on a stratified basis where matching is by age, sex, ethnic group, and no history of inhalant abuse in the individual or chronic inhalant abuse in his/her nuclear family. Data would be collected by means of a social history format that would utilize information already collected from schools, police, and social welfare, health, and other agencies. Furthermore, interviews with the subjects or collaterals or both should be undertaken. In this survey ethnographic information would also be collected (see Fredlund, 1994). Such a study should be piloted in one or two communities and, if feasible, carried out in others.

### End Stage Solvent/Inhalant Abusers

A pilot study is needed to seek and utilize a special sample of what are considered chronic adult inhalant/solvent abusers, including abusers in various communities. In this study, end stage would be defined as referring to those who have obvious and measurable mental and physical deficits believed to be a consequence of chronic inhalants or solvent use over many years. Investigation in this case would focus on individuals who are in their late twenties or older; in many cases, they would be individuals who have arrived at some sort of sheltered care or gravitated to homelessness. Variables and matching of experimental and control groups similar to those described in the case control study would be utilized. In this particular study the cooperation of collaterals (friends, family, and others) would be vitally important. Social history and ethnographic description would also be important. Such research might also provide cases for medical studies and neuropsychological assessment.

### *Surveillance of Inhalant Abuse Death*

In most states a majority of sudden and unexplained deaths are investigated by medical examiners. This procedure applies to New Mexico, where all unexplained deaths–about 35% of all deaths–are investigated. One might propose to undertake a retrospective study of all deaths that have come to the attention of the Office of the Medical Investigator that might be attributable to the use of inhalants or solvents (see Garriott & Petty, 1980). Furthermore, a prospective study design could monitor such deaths. Emphasis in such a study should be based on the physiology, as well as on the epidemiology, of these deaths. The epidemiology would focus on follow-back methods and consider a wide range of variables from psychosocial to toxicology for a more complete understanding than is currently presented in the literature. Much of the extant literature fails to link toxicology and medical considerations directly to behavioral patterns and psychological variables.

### *Fetal Solvent Syndrome*

Over the years some clinicians and researchers have suspected that there may be teratogenic effects on children born to heavy solvent/inhalant-abusing women. In fact, at least two articles were published raising the issue of whether there is a fetal gasoline or fetal solvent syndrome (Hunter, Thompson & Evans, 1979; Streicher et al. 1981). One might propose to identify communities and/or individuals suspected of having problems with fetal solvent and inhalant exposure. This study would then move to clinically examine these children by a dysmorphologist and identify any teratogenic effects that may be linked to or attributable to solvents and inhalants. By definition, this study would be quite small, but there are communities in various states and particular families in which such research could be pursued. The careful and very difficult sorting out of the teratogenic effects of polysubstance abuse would be vital to the methodology of this study. Similar defects may be produced by different drugs and interactive effects are also a challenge for analysis.

### *Longitudinal Cohort*

Although costly to conduct and demanding to implement with noncompliant or difficult-to-locate populations, a longitudinal design that tracks a cohort of users from initial vicarious use through the next stages of inhalant abuse, is necessary in order to identify factors involved in the progression from one stage of inhalant abuse to the next. Survey data are amenable to repeated measures analyses of multiple independent and dependent variables and would permit the development of a causal model. This approach would be dependent on a large initial sample size in order to have sufficient respondents from the original cohort involved in the later stages of the study. Large initial sample size is an important consideration as well, owing to the smaller prevalence rates for the later stages of inhalant abuse. The literature suggests the inclusion of a number of factors, such as family history of drug abuse, socioeconomic status (Bachrach & Sandler, 1985), disrupted family structure, peer associations, and social adjustment (Oetting et al. 1988), in order to explain or to predict the progression of inhalant abuse. The data collected would not be anonymous, but would be confidential; however, in order to collect valid self-report data, the researchers must establish some rapport with the respondents to encourage honest responses. Factors such as this add to the cost of a longitudinal design, but the gains in terms of the development of an explanatory model would outweigh the cost and difficulty of such a study.

## REFERENCES

Albaugh, B., & Albaugh, P. (1979). Alcoholism and substance sniffing among the Cheyenne and Arapaho Indians of Oklahoma. *International Journal of the Addictions, 14*(7), 1001-1007.

*Albuquerque Tribune* (1988). Gallup: A town under the influence (A special series reprint–originally appeared in September, 1988). Albuquerque, NM: Albuquerque Publishing Company.

Anderson, H. R., MacNair, R. S., & Ransay, J. D. (1985). Deaths from abuse of volatile substances: A national epidemiological study. *British Medical Journal, 290*(6464), 304-307.

Babst, D. V., Deren, S., Schmidler, J., Lipton, D. S., & Dembo, R. (1978). A study on family affinity and substance use. *Journal of Drug Education, 8*(1), 29-40.

Bachrach, K. M., & Sandler, I. N. (1985). A retrospective assessment of Indian abuse in the barrio: Implications for prevention. *International Journal of the Addictions, 20*(8), 1117-1189.

Barnes, G. E. (1980). *Northern Sniff: The Epidemiology of Drug Use among Indian, White, and Metis Adolescents.* Unpublished report, University of Manitoba, Departments of Psychiatry and Psychology.

Bass, M. (1970). Sudden sniffing death. *Journal of the American Medical Association, 212*(12), 1075-1079.

Beauvais, F. (1992a). Indian adolescent drug and alcohol use: Recent patterns and consequences. *American Indian and Alaska Native Mental Health Research, 5*(1), 1-67.

Beauvais, F. (1992b). Volatile solvent abuse: Trends and patterns. In C. W. Sharp, F. Beauvais, & R. Spence, (Eds.), *Inhalant abuse: A volatile research agenda* (NIDA Monograph No. 129, pp. 13-41). Rockville, MD: U.S. Public Health Service.

Beauvais, F., Oetting, E. R., & Edwards, R. W. (1985). Trends in the use of inhalants among American Indian adolescents. *White Cloud Journal, 3*(4), 3-11.

Chavez, E. L., & Swaim, R. C. (1992). An epidemiological comparison of Mexican American and White Non-Hispanic 8th and 12th grade students' substance abuse. *American Journal of Public Health, 82*(3), 445-447.

Cohen, S. (1977) Inhalant abuse: An overview of the problem. In C. W. Sharp, and M. L. Brehm, (Eds.), *Review of inhalants: Euphoria to dysfunction* (NIDA Monograph No. 15, pp. 2-11). Rockville, MD: U.S. Public Health Service.

Coulehan, J. L., Hirsch, W., Bullman, J., Sanandria, J., Welty, T. K., Colaiaco, P., Koros, A., & Lober, A. (1983). Gasoline sniffing and lead toxicity in Navajo adolescents. *Pediatrics, 71*(1), 113-117.

Crider, R. A., & Rouse, B. A. (Eds.). (1988). *Epidemiology of inhalant abuse: An update* (NIDA Monograph No. 85). Rockville, MD: U.S. Public Health Service.

De la Rosa, M., Khalsa, J. H., & Rouse, B. A. (1990). Hispanics and illicit drug use. *International Journal of the Addictions, 25*(6), 665-691.

Fredlund, E. V. (1994). *Volatile substance abuse among the Kickapoo people in Eagle Pass, Texas area, 1993.* Austin, TX: Texas Commission on Alcohol and Drug Abuse.

Garriott, J., & Petty, C. S. (1980). Death from inhalant abuse: Toxicological and pathological evaluation of 34 cases. *Clinical Toxicology, 16*(3), 305-315.

Guthrie, Patricia (1995). Stemming the ocean tide. *Albuquerque Tribune, 74*(54): A1-A6.

Hersey, C. O., & Miller, S. (1982). Solvent abuse: A shift to adults. *International Journal of the Addictions, 17*(6), 1085-1089.

Hunter, A., Thompson, D., & Evans, J. A. (1979). Is there a fetal gasoline syndrome? *Teratology, 20,* 75-80.

Jacobs, A. M., & Ghodse, A. H. (1987). Depression and solvent abusers. *Social Science and Medicine, 24*(10), 863-866.

Jacobs, A. M., & Ghodse, A. H. (1988). Delinquency and regular solvent abuse: An unfavorable combination? *British Journal of Addiction, 83,* 965-968.

Johnston, L. F., O'Malley, P. M., & Bachman, J. G. (1994). *National survey results on drug use from the Monitoring the Future Study, 1975-1993: Vol. 2. College Students and Young Adults.* Rockville, MD: National Institute on Drug Abuse.

Kauffman, A. (1973). Gasoline sniffing among children in a Pueblo Indian village. *Pediatrics, 51*(6), 1060-64.

Korman, M., Trumboli, F., & Semler, I. (1980). A comparative evaluation of 162 inhalant users. *Addictive Behaviors, 5,* 143-152.

Lalinic-Michaud, M., Subak, M., Ghadirian, A. M., & Kovess,V. (1991). Substance misuse among Native rural high school students in Quebec. *International Journal of the Addictions, 26*(9),1003-1012.

Liban, C. B., & Smart, R. G. (1982). Drinking and drug use among Ontario Indian students. *Drug and Alcohol Dependence, 9,* 161-171.

Mata, A. G., & Andrew, S. R. (1988). Inhalant abuse in a small, rural South Texas community: A social epidemiological overview. In R. A. Crider and B. A. Rouse (Eds.), *Epidemiology of inhalant abuse: An update* (NIDA Monograph No. 85, pp. 49-76). Rockville, MD: U.S. Public Health Service.

May, P. A. (1982). Substance abuse and American Indians: Prevalence and susceptibility. *International Journal of the Addictions, 17*(7), 1185-1209.

McSherry, T. M. (1988). Program experiences with the solvent abuser in Philadelphia. In R. A. Crider, and B. A. Rouse (Eds.), *Epidemiology of inhalant abuse: An update* (NIDA Monograph No. 85, pp. 106-120). Rockville, MD: U.S. PublicHealth Service.

National Institute on Drug Abuse. (1994). *National household survey on drug abuse: Population estimates, 1993.* Rockville, MD: U.S. Public Health Service.

Novak, A. (1980). The deliberate inhalation of volatile substances. *Journal of Psychedelic Drugs, 12*(2), 105-122.

Oetting, E. R., Edwards, R. W., & Beauvais, F. (1988). Social and psychological factors underlying inhalant abuse. In R. A. Crider and B. A. Rouse (Eds.), *Epidemiology of inhalant abuse: An update* (NIDA Monograph No. 85, pp. 172-203). Rockville, MD: U.S. Public Health Service.

Oetting, E. R., & Webb, J. (1992). Psychosocial characteristics and their links with inhalants: A research agenda. In E. R. Oetting and F. Beauvais (Eds.), *Inhalant abuse: A volatile research agenda* (NIDA Monograph No. 129, pp. 59-96). Rockville, MD: U.S. Public Health Service.

Padilla, E. R., Padilla, A. M., Morales, A., Olmedo, E. L., & Ramirez, R. (1979). Inhalant, marijuana, and alcohol use among barrio children and adolescents. *International Journal of the Addictions, 14*(7),945-964.

Reed, B. J. F., & May, P. A. (1984). Inhalant abuse and juvenile delinquency: A control study in Albuquerque, N.M. *International Journal of the Addictions, 19*(7), 789-803.

Remington, G., & Hoffman, B. F. (1984). Gas sniffing as a form of substance abuse. *Canadian Journal of Psychiatry, 29,* 31-35.

Sharp, C. W., Beauvais, F., & Spence, R. (Eds.). (1992). *Inhalant abuse: A volatile research agenda* (NIDA Monograph No. 129). Rockville, MD: U.S. Public Health Service.

Sharp, C. W., & Brehm, M. L. (Eds.). (1977). *Review of inhalants: Euphoria to dysfunction* (NIDA Research Monograph No. 15). Rockville, MD: U.S. Public Health Service.

Simons, J. F., & Kashani, J. (1980). Specific drug use and violence in delinquent boys. *American Journal of Drug and Alcohol Abuse, 7*(384), 305-322.

Smart, R. G. (1988). Inhalant use and abuse in Canada. In R. A. Crider and B. A.

Rouse (Eds.), *Epidemiology of inhalant abuse: An update* (NIDA Monograph No. 85, pp. 121-139). Rockville, MD: U.S. Public Health Service.

Streicher, H. Z., Gabow, P. A., Moss, A. H., Kono, D., & Kaehug, W. D. (1981). Syndromes of toluene sniffing in adults. *Annals of Internal Medicine, 94,* 758-762.

Swaim, R. C., Oetting, E. R., Thurman, P. J., Beauvais, F., & Edwards, R. W. (1993). American Indian adolescent drug use and socialization characteristics. *Journal of Cross-Cultural Psychology, 24*(1), 53-70.

Tinklenberg, J. R., Murphy, P., Murphy, P., & Pfefferbaum, A. (1981). Drugs and criminal assaults by adolescents: A replication study. *Journal of Psychoactive Drugs, 13*(3), 277-287.

Watson, J. M. (1980). Solvent abuse by children and young adults: A review. *British Journal of Addiction, 75,* 27-36.

Winfree, L. T., & Griffiths, C. T. (1983). Youth at risk: Marijuana use among Native American and Caucasian youths. *International Journal of the Addictions, 18*(1), 53-70.

Winfree, L. T., Griffiths, C. T., & Sellers, C. S. (1989). Social learning theory, drug use and American Indian youths: A cross-cultural test. *Justice Quarterly, 6*(3), 395-417.

Wingert, J. L., & Fifield, M. G. (1985). Characteristics of Native American users of inhalants. *International Journal of the Addictions, 20*(10), 1575-1582.

Zur, J., & Yule, W. (1990). Chronic solvent abuse: Relationships with depression. *Child: Care, Health, and Development, 16,* 21-34.

# Cultural Models of Inhalant Abuse Among Navajo Youth

Robert T. Trotter II, PhD
Jon E. Rolf, PhD
Julie A. Baldwin, PhD

**SUMMARY.** Drug abuse prevention among adolescents can be more effective if it is based on an accurate knowledge of the cultural context and of how young people actually think about the drugs that are commonly used. A study was undertaken among Navajo adolescents to query their perceptions of using drugs, what the social context of their drug use is and, in particular, their perceptions of inhalants as substances that are used to induce intoxication. *[Article copies available for a fee from The Haworth Document Delivery Service: 1-800-342-9678. E-mail address: getinfo@haworth.com]*

There is increasing concern over the extent of inhalant and other forms of substance abuse among American Indian school age children. Oetting, Goldstein and Garcia-Mason (1980) found that adolescents sampled from 5 different tribes showed a higher experimentation rate with drugs than adolescents from a national sample, comparing statistics for alcohol, inhalants, and illegal drugs. More recent studies (e.g., Beauvais et al. 1989) not

---

Robert T. Trotter II is Professor of Anthropology, Northern Arizona University. Jon E. Rolf is Professor of Health, Kansas State University. Julie A. Baldwin is Assistant Professor of Health Education, Northern Arizona University.

Address correspondence to: Robert T. Trotter II, Department of Anthropology, Campus Box 15200, Northern Arizona University, Flagstaff, AZ 86011.

[Haworth co-indexing entry note]: "Cultural Models of Inhalant Abuse Among Navajo Youth." Trotter, Robert T. II, Jon E. Rolf, and Julie A. Baldwin. Co-published simultaneously in *Drugs & Society* (The Haworth Press, Inc.) Vol. 10, No. 1/2, 1997, pp. 39-59; and: *Sociocultural Perspectives on Volatile Solvent Use* (ed: Fred Beauvais, and Joseph E. Trimble) Harrington Park Press, an imprint of The Haworth Press, Inc., 1997, pp. 39-59. Single or multiple copies of this article are available for a fee from The Haworth Document Delivery Service [1-800-342-9678, 9:00 a.m. - 5:00 p.m. (EST). E-mail address: getinfo@haworth.com].

 

only confirm these trends, they call for intervention strategies that begin in elementary school. As many as 22 percent of American Indian students may have used inhalants regularly (Coulehan et al. 1983; May, 1986). Our own survey of inhalant use among 8th grade Navajo students (1991-92) in the western part of the Navajo reservation, indicates that almost one fourth (24.4%) have tried inhalants at some time, and that 12.2 percent have used inhalants in the past month. The most frequently abused inhalants, such as gasoline, glue, and aerosols, are so accessible that they are the substances of choice for large numbers of American Indian children. This preference creates unique problems for drug abuse prevention, and suggests that it will be necessary to specifically target inhalants for special efforts. Unlike alcohol and illegal drugs, it is extremely difficult to even partially control access to these substances through legislation, and it is clearly impractical to attack their use through law enforcement efforts. Coincidentally, inhalants tend to be far more toxic (per use) than alcohol and other drugs. They cause physical damage at much lower doses and with far less length of use. In spite of these statistics and pragmatic conditions, the majority of drug prevention programs available to Indian youth focus on alcohol and illicit drugs, ignoring common household substances.

These special characteristics of inhalants create a serious need to systematically explore the nature of inhalant abuse among American Indian children to provide base line data on the types of inhalants used, the extent of their use, and to identify basic cultural themes that will be useful in assisting prevention efforts. This paper provides a model of inhalant use for one such group, the "Dine," or Navajo.

General substance abuse poses a substantial problem for the Navajo, as well as other American Indian groups. About 50% of American Indian youth are considered at high risk for some form of alcohol and drug abuse (Beauvais, Oetting & Edwards, 1985), with alcohol being the most commonly abused drug, and marijuana the second most common. However, for Navajo youth, there is growing concern by Tribal leaders and health officials over the abuse of inhalants. Therefore we have set a priority in this paper of providing evidence for the ways that inhalant abuse can be investigated within the context of the overall substance abuse problem of Navajo teenagers. It should be noted, that like alcohol and other drug use, there will be significant variation in inhalant use patterns for American Indian groups, based on geographic location, tribal affiliation, and within tribes (Heath 1985), especially since urban-rural substance abuse differences have been clearly documented (Weibel-Orlando, 1985; 1986/87), with abuse rates varying by group, setting, and for individuals between settings. Therefore, this paper presents baseline data for one group, which can then be tested in other groups before Pan-Tribal generalizations are made about inhalant abuse for other American Indian groups.

## *NAVAJO CULTURAL BACKGROUND*

The *Navajo,* or "Dine," are the largest American Indian tribe in the United States. Approximately 180,000 Navajo live on the Navajo Nation, while an additional 65,000 live outside the reservation in adjacent towns or in larger urban centers such as Denver, Los Angeles and Phoenix. The majority live on a 25,000 square mile reservation occupying the northeastern corner of Arizona and extending into parts of southern Utah and western New Mexico. The reservation is characterized by an arid climate and by the varied vegetation of the Colorado Plateau.

Navajo traditional culture involved pastoralism (primarily sheep and horses) from the 1800's well into the present. Living on scattered homesteads has been the norm for most of this century. The Navajo lifestyle has included an emphasis on the development of individual autonomy within the context of extended family units that can be counted on for assistance in time of need. Navajo philosophy embraces change wherever it can be incorporated into Navajo tradition. Consequently, the Navajo have modified both their language and their lifestyles through the generations in order to survive. In the recent past, the Navajo have adapted to the overall socioeconomic changes in the Southwestern United States by accepting new employment patterns and lifestyles, while continuing to maintain strong linguistic and cultural traditions. These economic changes have only marginally improved the socioeconomic condition of most Navajo. The isolation of the Navajo Nation, combined with high drop out rates and the need to travel off reservation for most employment opportunities, has severely restricted economic development for most tribal members. Unemployment rates on the reservation are extremely high at all times, and the overall level of income makes the area amongst the poorest in the United States.

The classic ethnographies of Kluckhohn and Leighton (1946) and Underhill (1956) offer comprehensive descriptions of the Navajo traditions, including historical sketches, notes on ceremonialism, language, and world view. Other literature covers specific facets of Navajo culture (e.g., Downs, 1972), a description of Navajo sheep herding (Reichard, 1974), treatise on religion (Witherspoon, 1975), and a detailed and extensive examinations of Navajo kinship and social structure (Lamphere, 1977). Correll's (1979) reexamination of Navajo history is noteworthy for its attempt to redress biases that have clouded European accounts of the

Navajo past. Overall, these works present a picture of a dynamic and adaptive traditional culture. While the Navajo have suffered from a long history of discrimination and neglect, their culture allows them to consistently succeed as an intact cultural group in the United States.

## *METHOD*

Implementing a culturally appropriate substance abuse program for Navajo teenagers requires a culturally competent understanding of their health beliefs, their models of substance abuse, and their culturally distinct behaviors. The authors have thus created a multidisciplinary and multimethod approach for exploring substance abuse risks for Navajo teenagers on the western portion of the Navajo reservation. The core of our project[1] is an in-school intervention program conducted as a sequence of in-class curriculum sessions in the 8th grade, followed a year later by a class program in the 9th grade. The curriculum is designed to improve American Indian teenagers' knowledge of HIV and other risks related to alcohol, drug abuse, and unsafe sexual practices. It is also designed to improve the teenagers' skills in avoiding or reducing risky behavior and to improve their skills in communicating with family, elders and peers. While much of this curriculum is not directly related to inhalant use, the lessons learned can be directly applied to any school based prevention effort, including inhalants.

This article reports findings from baseline ethnographic research obtained during the start up phase of the project. The ethnographic research concentrated on investigating three general cultural domains that were central to the development of the project curriculum. These are: (1) Navajo cultural models of adolescent alcohol use, (2) models of Navajo adolescents' use of other substances for getting high, and (3) their emerging attitudes and beliefs about HIV infection and AIDS. The majority of the data were collected during the first year of a four year collaborative research project. The research approach combines baseline quantitative data collection with ethnographic interviews, observations, and advanced ethnographic methods.

The data reported in this article are derived from open-ended, semi-structured focus groups and key informant interviews, supplemented by other systematic ethnographic methods (e.g., free listings, pile sorts, sentence frame completion, etc.). During the first six months of the project we conducted a total of 14 focus groups: three with adolescent females in three different high schools, three with adolescent males at the same schools, and eight with adults (Indian Health Service Health Advisory

Board members, dorm counselors, teachers, and parents), in the communities where the project was conducted. Two of the schools where interviews were conducted were in small towns located on the Navajo Reservation. The other school was in a nearby border town. In all but one instance, the moderator for each of the 14 focus groups was a Navajo interviewer. Female moderators facilitated the all female groups, while male moderators facilitated the all male groups. One of the male teenage groups was moderated by a Hispanic male who has field work experience with Native Americans locally. Each focus group utilized 4 focal questions, supplemented by six to eight major probes for each primary question. We also used multiple minor probes to assure comparable coverage of information between groups. The focus groups averaged two hours in length. They have been transcribed verbatim for computer based ethnographic analysis.

The transcripts were initially created as WordPerfect files and converted to ASCII files. The textual files were analyzed with the assistance of TALLY 3.0, which is a text-based content and ethnographic analysis program. Names have not been used in this article to protect the confidentiality of our informants. The data were analyzed using a multi-level coding scheme which addressed cultural domains, content, and processes of interest to researchers (*a priori* codes), as well as content areas which were embedded in the text itself (emic codes). Analysis and interpretation of the data from the transcripts were carried out collectively by the investigators with the help of community representatives. Subsequent analysis utilizing other software programs has allowed us to accomplish content analysis, domain and thematic evaluation, multidimensional scaling and cluster analysis of ethnographic data sets. These programs include CONCORD (a concordance program) and ANTHROPAC 4.0 (a data processing package for systematic ethnographic data analysis).

In addition to qualitative interviewing, the authors created free-listing exercises embedded in a questionnaire with open and closed-ended questions, which were administered just prior to the focus group interviews, or to representative groups at other times. The free-listing exercise uses questions that allow people to provide an unconstrained list of all the things that are important about a particular cultural domain (cf., Bernard, 1988; Weller & Romney, 1988). These questions act as a preliminary example of the types and content of the questions that will be addressed by the focus groups, and they provide an opportunity for privately expressing an opinion on the topic.

This technique permitted the use of questions that were specifically designed to elicit ideas, knowledge, and opinions about events while also protecting Navajo cultural values. This approach can also reduce the distress members of a group might feel about discussing intimate topics by giving them the opportunity to think about the questions in advance. It also allowed us to cross check some of the answers given privately with each individual's responses in the focus group session.

In order to create a comparative data base for the free listings from Navajo adolescents and adults, we conducted two sessions of free listings on "alcoholic beverages" and "things, other than alcohol, that people use to get high" with American Indian and Anglo[2] students from three introductory anthropology classes at Northern Arizona University. This data provided comparative data which allowed us to check the range and rank order of the free listings provided by the Navajo youth and adults.

The advantage of utilizing the American Indian college students as a comparative base is that they are living in the same sociocultural region, are fairly close in age to the Navajo teenagers, have similar access to alcohol and drug resources, and experience similar media exposure to alcohol, drugs and AIDS. As college students, they are older and have more education than the high school students, which may cause them to differ in terms of motivation, orientation toward success, educational expectations, and perhaps even family values. These variables, however, are not directly at issue in developing simple free listings on alcohol and drugs.

## *RESULTS*

The results of our ethnographic interviews and free listings are organized into two sections: cultural themes relevant to prevention efforts and models of inhalant abuse.

### *General Cultural Themes*

Our research objectives were predominantly action oriented; they were aimed at enhancing the project's curriculum content, producing evaluation instruments, and providing process evaluation information. We asked Navajo teenagers and adults to identify the cultural barriers or the resistance points that need to be overcome to change substance abuse problems. Their recommendations indicate that there are at least four primary cultural themes that confirm a need for a culturally sensitive approach to drug prevention efforts for Navajo adolescents. The first theme is *the high level of positive regard for individual autonomy expressed in Navajo culture*; the second is *the existence of a serious generation gap between Navajo youth and adults*; the third is *a general respect for authority*, and the fourth is *the strength of cultural norms about modesty in relation to the discussion of intimate topics by Navajo.*

*Autonomy.* There is a strong Navajo cultural value to respect the autonomy of the individual. This is most commonly expressed in the comment that once someone has been given proper knowledge or information, then what that person does with the information is "up to them." Over the course of the focus group sessions this attitude manifested itself on a number of occasions and clearly effects the way in which inhalant abuse prevention and education must be approached. Navajo teenage boy commenting on intervention in an alcohol problem:

> If I was trying to help a friend it depends up to him, if he wanted to stop he can stop. I can help. If he doesn't want to, his own choice. Can't do nothing about it.

Navajo teenage girl discussing preventing risks:

> Well, um, . . . [you] can tell [them] about things, perhaps you could encourage them to get seen, but that would fall back on the individual themselves.

*The Generation Gap.* One of the purposes of the current project is to identify the proper channels of communication for prevention messages directed at youth and adults in Navajo communities. This is no simple task because the program deals with sensitive topics, especially sexual issues. Navajo youth, as a rule, have a great deal of respect for their elders and for knowledgeable authorities. It may be ironic that this respect can produce difficulty in communication about risky behavior or problems in school. When we asked, "What member in your family, your relatives, would you talk with the most about difficult subjects?" the teenagers replied that they would talk to their sisters, cousins and brothers more often than they would talk to anyone else. Another family member that was frequently mentioned was their aunt. This resistance to talking with most adults is similar to that found in other adolescent populations. A preference for seeking advice from siblings has been noted in many other studies. Navajo parents were occasionally asked for advice about intimate subjects such as sex, drugs, difficulties in school, or other teenage problems, but more commonly they were not approached.

> I'm more comfortable talking with my sister about my personal stuff, and you're asking me who I'm not very comfortable with . . . ?

> My mom. every time I ask her something she say "you're not grown up enough to ask that question." (Navajo Teenage Girl)

The following response is typical of the teenagers who have difficulty in communicating with their parents.

> Some things are hard to tell your parents 'cause you never know what they could do to you. Some things they'll say, like if you tell them you have done this, just tried it, they'll get real mad. I've known some girls' parents are like that. (Navajo Teenage Girl)

While this generation gap persists, there is also a countervailing trend of keeping intimate conversations and details within the extended family.

*Respect for Authority.* Knowledge and experience are generally respected by Navajo teenagers. They respond to the authority of publicly acknowledged leaders (elders, physicians, teachers, etc.), at least in terms of the people they say that they would seek to advise them about drug abuse. The Navajo kinship pattern is a matrilineal descent system, in which membership in and identification with a particular clan is a very important social process, especially for the more traditional families. This means that the primary individuals who would be parent surrogates for drug abuse intervention and any type of family based prevention discussions, would normally be people who were a part of these extended family systems. The following exchange demonstrates that while parents are difficult to approach, there is a preference for talking with trusted seniors, especially within the extended family. When asked who the students could talk to, one replied:

> Respondent: The one who understands when you talk to them. The one you're most like comfortable talking to. The one mostly you really get along with.
>
> Interviewer: So does it matter on age?
>
> Respondent: Yeah.
>
> Interviewer: What age do you think you'd feel more comfortable with? Same age, or someone older?
>
> Respondent: Someone older.

*Modesty and Discussions of Sex or Other Intimacies.* Among Navajo traditional people, intimate topics are not openly discussed. Adolescents feel serious discomfort in talking about alcoholism, drug abuse and AIDS with their parents, especially if the parents are traditional. In addition, the results of our adult focus groups indicate that parents, even if they are health professionals, often have an equally difficult time discussing sexuality and AIDS with their children.

The focus groups and other interview methods all indicate that sex was the most difficult subject for individuals to discuss. Alcohol topics were easiest to approach, other drug topics a little harder, and sexual subjects the most difficult of all. Discussion of sex was avoided by both adults and teenagers, and by both males and females, with somewhat less reticence evident among the teenage girls. Every time we first mentioned the subject in focus group interviews, everyone stared at the floor and did not say anything for quite a while. Fortunately, our interviewers were able to develop sufficient rapport to eventually gain information in this area. The following interchange was similar to our overall experiences in exploring any subject that dealt with sexual activities:

> Respondent: Teenagers, it's really hard for them to talk about drugs, sex and alcohol, bring it up with any parent or any adult.
>
> Interviewer: Okay, let's say you want to talk about alcohol with your parent or your guardian. What makes it hard for you to talk about them?
>
> Respondent: Too embarrassed.

In our adult focus groups there was overwhelming consensus that schools, not the parents and certainly not the traditional parents, were the proper forum for sex education. They felt that the parents simply could not overcome their modesty to discuss this subject effectively. In many cases, this recommendation extended to other intimate topics, such as substance abuse, family problems, school problems, and the like. This condition suggests that the traditional parents are not the appropriate target for recommendations for behavioral changes in at least some of these areas, contrary to the recommendations of the more conservative political elements in the United States.

### *Cultural Models of Substance Abuse by Navajo Adolescents*

We divided our investigations of the teenager's exposure to substance abuse into two complimentary research foci. The first was an exploration of their knowledge, beliefs, and behaviors relating to alcohol use, and the other was their cultural models of other types of substance abuse. This division overlaps with the Navajo teenager's own discussion of substance abuse problems but is also somewhat different from it. The teenagers discuss alcohol related problems as one clearly defined area of substance abuse and tended to lump the other types of abuse together. However, it is also clear from our analysis of the focus group, and other ethnographic data, that these adolescents sub-divide "other substance abuse" into illegal drugs and inhalants. They discuss inhalant abuse as a very important third category of substance abuse that is highly prevalent on the reservation.

*Adolescent and Adult Views of Reservation Inhalant Use.* Navajo adolescents experiment with a wide variety of substances that can make them "high." Our adolescent respondents mentioned more names for "drugs," and more types of substances that their peers used than types and brand names of alcoholic beverages. The free listing in Table 1 "drugs" was created by requesting our informants to list all of the drugs, except alcohol, used by Navajo youth. This exercise provides a good example of the prevalence of inhalants in the drug repertoire of these teenagers.

The importance of inhalants in the overall drug abuse exposure of Navajo teenagers is obvious from this listing. A total of 15 of the listed substances were inhalants (34.9% of the items listed). Equally important is that inhalants constitute more than 50 percent of the ten most frequently named substances, especially when three of the top ten are synonyms for marijuana (marijuana, pot, weed).[3] Since a free listing produces the most salient items in a cultural domain, the common inhalants identified as being used by these students are petroleum products (gasoline and kerosene), glue, white out, nail polish, and paint products.

We collected a parallel free listing data set from the parents of some of the teenagers, as well as some of their teachers and local Navajo health officials (Table 2). This data set provides a complimentary view of substance abuse for Navajo youth. The results provide an interesting triangulation of the types of drugs that the teenagers are exposed to on the reservation. The results indicate that the adults see the prevalence of inhalant abuse in much the same way as the teenagers themselves.

The adult drug listing items include 37.0 percent that are inhalants, including the top 5 items listed. The most commonly used substances in the two groups show significant overlap, with the most salient substances listed by adults being petroleum products, white out, glue, hair spray, and paint products.

We ran an additional comparison on the listings given by students who lived on the reservations and those that lived in nearby border towns

TABLE 1. Free Listing of Drugs Used by Navajo Teenage Respondents

| ITEM | | FREQUENCY | RESP PCT |
|---|---|---|---|
| 1 | MARIJUANA | 12 | 46 |
| 2 | COCAINE | 10 | 38 |
| 3 | GASOLINE | 7 | 27 |
| 4 | WEED | 5 | 19 |
| 5 | GLUE | 4 | 15 |
| 6 | NAIL POLISH | 4 | 15 |
| 7 | WHITE OUT | 3 | 12 |
| 8 | HAIR SPRAY | 3 | 12 |
| 9 | ALCOHOL | 3 | 12 |
| 10 | POT | 3 | 12 |
| 11 | DON'T KNOW | 3 | 12 |
| 12 | ACID | 2 | 8 |
| 13 | LSD | 2 | 8 |
| 14 | COKE | 2 | 8 |
| 15 | PILLS | 2 | 8 |
| 16 | ANGEL DUST | 2 | 8 |
| 17 | HEROIN | 2 | 8 |
| 18 | CRACK | 2 | 8 |
| 19 | CRYSTAL | 2 | 8 |
| 20 | CIGARETTES | 2 | 8 |
| 21 | KEROSENE | 2 | 8 |
| 22 | OIL | 2 | 8 |
| 23 | PAINT | 2 | 8 |
| 24 | RUBBER CEMENT | 2 | 8 |
| 25 | SPRAY PAINT | 1 | 4 |
| 26 | MAGIC MARKERS | 1 | 4 |
| 27 | UNDER ARM DEODORANT | 1 | 4 |
| 28 | SUPER GLUE | 1 | 4 |
| 29 | ANYTHING THAT SMELLS STRONG | 1 | 4 |
| 30 | SLEEPING PILLS | 1 | 4 |
| 31 | SPEED | 1 | 4 |
| 32 | SNIFF | 1 | 4 |
| 33 | ROACH | 1 | 4 |
| 34 | PEYOTE | 1 | 4 |
| 35 | DOWNERS | 1 | 4 |
| 36 | SKOAL | 1 | 4 |
| 37 | HALLUCINOGENS | 1 | 4 |
| 38 | PCP | 1 | 4 |
| 39 | GRASS | 1 | 4 |
| 40 | TOBACCO | 1 | 4 |
| 41 | RUBBING ALCOHOL | 1 | 4 |
| 42 | DRUGS | 1 | 4 |
| 43 | UPPERS | 1 | 4 |
| Total mentions; mentions per respondent | | 102 | 3.92 |

N = 26 (12 male, 14 female)

TABLE 2. Free Listing of Drugs Used by Navajo Adolescents According to Navajo Adult Respondents

| ITEM | | FREQUENCY | RESP PCT |
|---|---|---|---|
| 1 | WHITE OUT | 7 | 41 |
| 2 | PAINTS | 6 | 35 |
| 3 | HAIR SPRAY | 4 | 24 |
| 4 | GASOLINE | 4 | 24 |
| 5 | GLUE | 4 | 24 |
| 6 | NO RESPONSE | 3 | 18 |
| 7 | NAIL POLISH | 3 | 18 |
| 8 | MAGIC MARKER | 3 | 18 |
| 9 | COCAINE | 2 | 12 |
| 10 | MARIJUANA | 2 | 12 |
| 11 | SPRAY PAINT | 2 | 12 |
| 12 | DON'T KNOW | 2 | 12 |
| 13 | HEROIN | 1 | 6 |
| 14 | CODEINE | 1 | 6 |
| 15 | COKE | 1 | 6 |
| 16 | WEED | 1 | 6 |
| 17 | INK | 1 | 6 |
| 18 | CIGARETTE | 1 | 6 |
| 19 | TOBACCO | 1 | 6 |
| 20 | SCOUT | 1 | 6 |
| 21 | LOVEWEAKER | 1 | 6 |
| 22 | COUGH SYRUP | 1 | 6 |
| 23 | NONE | 1 | 6 |
| 24 | POT | 1 | 6 |
| 25 | MOUTHWASH | 1 | 6 |
| 26 | RUBBING ALCOHOL | 1 | 6 |
| 27 | CROP OUTS | 1 | 6 |
| Mentions; mentions per respondent | | 57 | 3.35 |

N = 17 (males = 6, females = 11)

(comparing the two reservation schools with the border town school). There were no significant differences in terms of the types of drugs listed when comparing the border town students with the adolescents from reservation schools. We originally hypothesized that greater availability of illegal drugs in the border towns might have an effect on the most frequently mentioned items (assuming a greater variety of drugs being available, and more being known in the border towns), which could result in a differential in the number of responses per student and the total variety of responses by location. This hypothesized differential was not supported by the data.

*Comparative Data from College Students.* We collected a parallel data set from Navajo college students, using the free listing technique (Table 3).

However, through a serendipitous mistake, the question we used with the college students was, "please name all of the things, except alcohol, you know of that people use to get high," as opposed to the question we used with the school students, which was "please name all of the drugs, except alcohol, that people use to get high."

The Navajo college student free listings provide several contrasts with the high school data. There appears to be a significant increase in these students' awareness of the different things that make people high. They provide an average of almost 12 items in each of their lists, compared with an average of about 4 items per respondent in the high school students' lists. The importance of inhalants is diminished in this group, representing only 20.5 percent of the total items mentioned, although some new items, such as Xerox correction fluid and finger nail polish remover, are added to the overall list. The college student list has only 3 inhalants in the top ten listings as opposed to the higher percentage for the high school students and the adults. The listings from the Anglo college students, not presented here, list even fewer inhalants and, as one consistent cross cultural difference, include frequent mention of nitrous oxide, laughing gas (from whipping cream cartridges), which is missing from all of the American Indian free-listings.

At least some of these differences were probably produced by the different wording in the basic question for the free listing task. Due to the more generic wording, the college version of the question produced a list of behaviors, in addition to substances, as ways of getting high. These behaviors include sex, exercise, thrill seeking, and mystic experiences, some of which could be considered as possible positive alternatives to inhalants and drugs. They represent 22.9 percent of the total items listed. Without their inclusion (assuming the same lists minus these behaviors), inhalants would have accounted for 26.4 percent of the listings, which is lower than but closer to the percentages suggested by the high school students.

The change in wording was accidental, resulting from a need to immediately get a scan of drugs being used, but the difference in responses is salient from two perspectives. First, the broader wording provides a more comprehensive and realistic inventory of the risks that are found in this community. Second, it includes some conditions that do not necessarily have socially negative consequences. One notable gender difference in this data is that more females than males provided the non-drug behavioral responses in the college student free listings. We are following up on the implications of their answers.

This list also includes several examples of Navajo humor as exemplified by the answers "elevators" and "flying," as things that get a person high. This type of joking and putting-on has been common throughout our

TABLE 3. Free Listing of All Things People Use to Get High, Except Alcohol by American Indian College Students

| ITEM | | FREQUENCY | RESP PCT |
|---|---|---|---|
| 1 | PEYOTE | 10 | 59 |
| 2 | CRACK | 10 | 59 |
| 3 | COCAINE | 10 | 59 |
| 4 | MARIJUANA | 9 | 53 |
| 5 | GLUE | 9 | 53 |
| 6 | ACID | 8 | 47 |
| 7 | GASOLINE | 8 | 47 |
| 8 | COKE | 7 | 41 |
| 9 | PAINT | 7 | 41 |
| 10 | HEROIN | 6 | 35 |
| 11 | LSD | 5 | 29 |
| 12 | CIGARETTES | 5 | 29 |
| 13 | SPEED | 5 | 29 |
| 14 | POT | 5 | 29 |
| 15 | MUSHROOMS | 4 | 24 |
| 16 | WEED | 4 | 24 |
| 17 | SEX | 4 | 24 |
| 18 | HAIR SPRAY | 3 | 18 |
| 19 | PCP | 3 | 18 |
| 20 | PAINT THINNER | 3 | 18 |
| 21 | WHITE OUT | 3 | 18 |
| 22 | CAFFEINE | 3 | 18 |
| 23 | DOWNERS | 3 | 18 |
| 24 | UPPERS | 3 | 18 |
| 25 | OVER THE COUNTER DRUGS | 2 | 12 |
| 26 | SPRAY PAINT | 2 | 12 |
| 27 | PAINT AND SOCKS | 2 | 12 |
| 28 | ANGEL DUST | 2 | 12 |
| 29 | HASH | 2 | 12 |
| 30 | LISTERINE | 2 | 12 |
| 31 | CRYSTAL | 1 | 6 |
| 32 | STIMULANTS | 1 | 6 |
| 33 | ORAL MEDICATION | 1 | 6 |
| 34 | PRESCRIPTION DRUGS | 1 | 6 |
| 35 | HALLUCINOGENS | 1 | 6 |
| 36 | JOINT | 1 | 6 |
| 37 | FINGER NAIL POLISH REMOVER | 1 | 6 |
| 38 | XEROX CORRECTION FLUID | 1 | 6 |
| 39 | NATURAL HIGH | 1 | 6 |
| 40 | DOPE | 1 | 6 |
| 41 | TOBACCO | 1 | 6 |
| 42 | MONEY | 1 | 6 |
| 43 | SPRAYS | 1 | 6 |
| 44 | ELEVATOR | 1 | 6 |
| 45 | FLYING | 1 | 6 |
| 46 | BIRTH OF BABY | 1 | 6 |
| 47 | SKY DIVING | 1 | 6 |
| 48 | ANYTHING THRILLING | 1 | 6 |

| ITEM | | FREQUENCY | RESP PCT |
|---|---|---|---|
| 49 | ANYTHING DARING | 1 | 6 |
| 50 | CLEANING PRODUCTS | 1 | 6 |
| 51 | AA MEETINGS | 1 | 6 |
| 52 | COMBINATIONS | 1 | 6 |
| 53 | ICE | 1 | 6 |
| 54 | JOY RIDING | 1 | 6 |
| 55 | CAR RACING | 1 | 6 |
| 56 | ASPIRIN | 1 | 6 |
| 57 | INDULGING IN FOOD | 1 | 6 |
| 58 | IV DRUGS | 1 | 6 |
| 59 | SHAVING LOTION | 1 | 6 |
| 60 | SHERM SMOKED WITH PCP | 1 | 6 |
| 61 | DEPRESSANTS | 1 | 6 |
| 62 | COOL BREEZE | 1 | 6 |
| 63 | HANG UPSIDE DOWN | 1 | 6 |
| 64 | SELF SATISFACTION | 1 | 6 |
| 65 | DRUGS | 1 | 6 |
| 66 | NAIL POLISH | 1 | 6 |
| 67 | EXERCISE | 1 | 6 |
| 68 | QUAALUDES | 1 | 6 |
| 69 | FREEBASE | 1 | 6 |
| 70 | MUSIC | 1 | 6 |
| 71 | JOGGING | 1 | 6 |
| 72 | OPIUM | 1 | 6 |
| 73 | KEROSENE | 1 | 6 |
| 74 | TOBACCO JUICE | 1 | 6 |
| 75 | RUBBER CEMENT | 1 | 6 |
| 76 | SELF RIGHTEOUSNESS | 1 | 6 |
| 77 | PILLS | 1 | 6 |
| 78 | ECSTASY | 1 | 6 |
| 79 | AEROSOL SPRAY CAN | 1 | 6 |
| 80 | PURE VANILLA | 1 | 6 |
| 81 | SOBRIETY | 1 | 6 |
| 82 | ANYTHING TOXIC | 1 | 6 |
| 83 | INHALANTS | 1 | 6 |
| Total mentions; mentions per respondent | | 202 | 11.882 |

N = 17 (2 male, 15 female)

interviews, and it is an important element in the relationship between the teenagers and anyone who wants to have effective interactions with them. It is a reminder that it is important to keep one's sense of humor when providing educational materials to any group of teenagers, and to Navajo teenagers in particular.

## *Focus Group Data on Other Substance Use and Abuse*

We discovered three basic elements in a Navajo teenage cultural model of inhalant abuse which can be presented through the excerpts from focus group transcripts created by our questions to Navajo teenagers. We explored the types of drugs in use, the reasons for their use, and the consequences of use as seen by adolescents. All of these issues were relevant to developing our prevention programs. We felt it was important to allow the teenagers to tell us which factors were present on the Navajo reservation and what the level of importance for each factor was, from their perspective instead of ours.

*Kinds of Drugs in Use on the Reservation.* Our focus group interviews produced a very long list of drugs in use, which closely parallel the data from our separate free listing data collection. The following is a summary listing created by one of the teenage focus groups when they were asked to name all of the drugs in use in their schools and communities:

> Alcohol, marijuana, pills, inhalants, glue, hash, acid, ice, spray can, gasoline, white out, nail polish, listerine, speed, dope, weeds, peyote, heroin, PCP, LSD, uppers, downers, angel dust, homemade drugs, inhalants, mouthwash, speedballs, rubber cement, spray paint, lysol, crack crystal, hair spray, nitrous oxide, anything that has alcohol in it.

Inhalants, once again, are among the most commonly listed drugs that students use to get high, representing 28.6 percent of the items in this consensual list. The others break down into alcohol abuse (alcohol, listerine, mouthwash, and anything that has alcohol in it) and illicit drugs. Older teenagers attribute the highest use of inhalants to the younger teens. There is an incredible variety of inhalants available, most of which are everyday household items. Inhalant abuse ranges from putting gasoline or spray paint in milk cartons and closing the top of the carton around your mouth and nose to inhaling spray paint that is imbedded in socks. When asked about people using some of these household intoxicants, we received the following information during our focus group interviews.

Navajo male on spray paint:

> Yeah. He had one of those, and he put some of that spray paint in that carton of milk and he was getting high on it.

It also appears that inhalant use can sometimes be combined with drinking items from the aerosols, after sniffing them. The liquid left over after the propellant is gone may contain alcohol or another intoxicant.

Navajo female on drinking hair spray:

> And now the people on the Res are starting to drink hair spray. They poke the bottom of it, I saw a couple of friends in (reservation town),

> and you poke the bottom of it and the air all inflates out and then they take the top off. I don't know how they took it off. And they just drinked [sic] it. They took, like, 5 bottles of Aquanet. In an hour they all passed out.

One difference that we encountered regarding the use of inhalants, as compared to alcohol use, and to some extent drug use, is that the teenagers had very definite negative views of the "sniffers" compared with other users. They were considered to look and even smell different than the other kids. Their eyes were kind of "crazy" or weird looking, and they often were thought to have physical evidence of abuse such as paint around their nose or mouth, or the distinct smell of gasoline on their clothes. If there is a hierarchy of drug users among Navajo teenagers, the ones who do inhalants regularly are thought to be at the bottom of the social ladder. For prevention efforts, this means that these individuals tend to be marginalized and consequently are harder to reach, may be less effected by changing the norms of their peer group, and may have far less social support and reason to change their behavior than other teenagers.

*Reasons for Using Drugs.* The reasons the teenagers gave for using drugs did not differ significantly when they were talking specifically about inhalants as opposed to other types of drugs. The most common reasons were to stay with their friends, because of curiosity, to escape unhappiness or harsh realities, to get high and forget problems, and to feel good. One teenager gave the following example that includes these reasons:

> They probably like how they feel when they're high, like, all laughing and giggling. I have a friend who does that, when she gets high all I hear is laughing and giggling. I guess they like how they feel.

The teenagers also talked of doing drugs because of older relatives, especially siblings, doing them, and of wanting to stay in a particular group of friends. They also commonly added "being cool" as one of the reasons for doing drugs.

> Um. You do it just to feel good and be cool, and once you do the drugs you get addicted.

*The Consequences of Drug Use.* The teenagers recognize that there are serious consequences for substance abuse. When we asked if people were more likely to do dangerous or risky things while they are high, the students not only came up with types of things that people do, they gave specific examples from their own family and friends. Some of the things they mentioned included jumping from cliffs, crashing a car, suicide, killing relatives while they were high, other forms of interpersonal violence, and acting crazy. They also included accidents in their list. One teenager told the story of a younger sister who was into gasoline sniffing. She would walk around with her nose in a milk carton, to stay high. One day she walked out into a road and was killed by a passing car. Virtually everyone else in the focus group could recount a similar story about a relative or someone they knew, with the inhalant varying from paint to kerosene to glue. Not everyone was killed in these stories, but all of them suffered serious accidents. Yet, while they recognize the consequences of inhalant and other drug use, many still become involved with this type of risk.

## *CONCLUSIONS*

We have begun to describe the model of adolescent inhalant abuse that is present on the Western Agency of the Navajo Nation. These teenagers have a somewhat differing but overlapping cultural model of substance abuse when compared with Anglo American teenagers. The responses of the Navajo teenagers are similar to other adolescents in such issues as having difficulty in talking with parents and their response to peer pressure (c.f., Chilman, 1983; Offer, 1986; and Hayes, 1987). Navajo teenagers' responses differ in other ways, however, especially concerning the importance of the autonomy of the individual, the importance of inhalants as drugs, the importance of the extended family in determining positive and negative behaviors, and trouble talking about sex due to specific Navajo cultural prohibitions. The Navajo inhalant abuse model includes a realistic view of the processes and consequences of inhalant use, coupled with a continuation of deleterious consumption patterns for some youth.

The cultural information provided by Navajo youth and adults has been incorporated into our in-school prevention program. This includes recognition of the importance of individual autonomy as well as the extended family in Navajo tradition and modern life. The research has also pointed out the importance of incorporating Navajo traditional beliefs and values in the revised curriculum. We feel that school-based prevention programs can become more powerful if they are effectively linked with family practices and values that are familiar and culturally congruent for the students. These beliefs and practices include using appropriate communication channels incorporating both positive and negative modeling from parents, siblings, and extended family members.

The risk prevention package for school-aged youth combines the following components which are adapted for cultural appropriateness: (1) teen-parent/parent-surrogate communications to encourage preventive attitudes in extended family contexts, (2) community-directed communication to adults, tribal elders and opinion leaders for acceptance of prevention programs, (3) educational curricula for school contexts, and (4) social skill training for peer contexts. Some of the social skill training is generic and would be appropriate for any teenage population. It emphasizes self-efficacy by providing opportunities for students to observe others effectively responding to pressure situations and by reinforcing resistance skills through role playing exercises.

This entire process of "localizing" prevention efforts through the type of research effort presented in this paper, combined with applicable psychosocial and cultural theories, appears to be highly desirable. Our evaluation strategies indicate that this type of intervention is successful in increasing the level of knowledge about risks and reducing risk-taking behavior among Navajo teenagers on the reservation.

## NOTES

1. The project from which the data was partially derived, the Native American Prevention Project of AIDS and Substance Abuse (NAPPSASA), funded by and NIAAA grant R01-AA08578 (J. Rolf and C. Alexander, P.I.'s) is a preventive intervention study involving school and community programs. It is designed to reduce the future incidents of HIV infection, alcoholism and drug abuse in American Indian adolescents. The project's prevention programs are being developed and evaluated by combining culturally sensitive, qualitative and quantitative research methods applied by a consortium of persons from Johns Hopkins University Department of Maternal and Child Health and Northern Arizona University's Department of Anthropology, in cooperation with Navajo experts from public schools and community organizations. Following common anthropological ethical guidelines, the schools, community organizations and individuals are not being identified, to protect the confidentiality preferences of our informants. However, it should be noted that this project was conducted on the Western Agency of the Navajo Nation and that the results should not be taken as necessarily representative of the Navajo Nation as a whole, since there are well-documented linguistic, cultural, social and economic variation between regions of the Navajo Nation.

2. We are using the term "Anglo" congruently with the Southwestern U.S. practice of labeling people Anglo if they are members of the dominant U.S. cultural system and are not Hispanic, American Indian, African American or Asian.

3. The synonyms found in the free listing exercise (e.g., marijuana, pot, roach and weed) have been deliberately combined into a single category. They have been kept separate where the same respondent provided more than one label for the substances being used and where combining them would give a false impres-

sion of the percentage of informants who mentioned a particular substance. We also felt that it was important to keep the full range of variation in the listing in order to identify the regional linguistic variation in the names of these substances. These variants can then be used in localizing prevention and education efforts, using the adolescents' own terminology for these substances.

## REFERENCES


Beauvais, F., Oetting, E. R., and Edwards, R. W. (1985). Trends in drug use of Indian adolescents living on reservations. *American Journal of Drug and Alcohol Abuse,* 11(4):1975-1983.

Beauvais, F., E. R. Oetting, W. Wolf, and R. W. Edwards (1989). Native American Youth and Drugs, 1976-87: A Continuing Problem. The *American Journal of Public Health,* 79(5):634-636.

Bernard, H. R. (1988). *Research Methods in Cultural Anthropology,* Beverly Hills. Sage.

Chilman, C. (1983). *Adolescent Sexuality in a Changing American Society.* 2nd Ed., New York: John Wiley and Sons.

Correll, J. L., ed. (1979). *Through White Man's Eyes; A Contribution To Navajo History: A Chronological Record of the Navajo People from Earliest Times to the Treaty of June 1, 1868.* Window Rock: Navajo Heritage Center.

Coulehan, J. L., Hirsch, W., Brillman, J., Sanandria, J., Welty, T. K., Colaiaco, P., Koros, A., & Lober, A. (1983). Gasoline sniffing and lead toxicity in Navajo adolescents. *Pediatrics,* 71(1):113-7.

Downs, J. F. (1972). *The Navajo.* New York. Holt, Rinehart and Winston.

Hayes, C. (ed.) (1987). *Risking Our Future: Adolescent Sexuality, Pregnancy and Child Bearing.* Vol I. Washington, D.C. National Academy Press.

Heath, D. B. (1985). Native Americans and Alcohol: Epidemiological and Social Relevance. In *Alcohol Use Among U.S. Ethnic Minorities,* NIAAA Research Monograph No. 18, DHHS No. (ADM) 89-1435, Rockville, MD, pp. 207-222.

Kluckhohn, C. and D. C. Leighton (1946). *The Navajo.* Cambridge, MA: Harvard University Press.

Lamphere, L. (1977). *To Run After Them: Cultural and Social Bases of Cooperation in a Navajo Community.* Tucson: University of Arizona Press.

May, P. A. (1986). Alcohol and drug misuse prevention programs and Native Americans: Needs and opportunities. *Journal of Studies on Alcohol,* 47(3):187-95.

Oetting, E. R., Goldstein, G. S., & Garcia-Mason, V. (1980). Drug use among adolescents of five southwestern Native American tribes. *International Journal of Addictions,* 15:439-445.

Offer, D. (1986). "Adolescent Development: A Normative Perspective." In A. Frances and R. Hales (eds.) Annual Review, Vol 5, Washington D.C.: American Psychological Association, pp. 404-419.

Reichard, G. (1974). *Navajo Religion: A Study of Symbolism. Princeton*: Princeton University Press.

Underhill, R. (1956). *The Navajos.* Norman: The University of Oklahoma Press.
Weibel-Orlando, J. (1985). *Indians, Ethnicity, and Alcohol: Contrasting Perceptions of the Ethnic Self and Alcohol Use.* In Linda A. Bennett and Genevieve Ames, American Experience With Alcohol, New York: Plenum, pp. 201-226.
Weibel-Orlando, J. (1986/87). Drinking patterns of urban and rural American Indians. *Alcohol Health & Research World,* 2(2):8-13.
Weller, S. and A. K. Romney (1988). *Systematic Data Collection.* Newbury Park, CA: Sage Publications
Witherspoon, G. (1975). *Navajo Kinship and Marriage.* Chicago: The University of Chicago Press

# Volatile Solvent Use: Patterns by Gender and Ethnicity Among School Attenders and Dropouts

Scott C. Bates, MS
Bradford Wong Plemons, MS
Pamela Jumper-Thurman, PhD
Fred Beauvais, PhD

**SUMMARY.** Differences in patterns of volatile solvent use were explored with special emphasis on use as related to school enrollment status. The sample included American Indian, Mexican-American and White American youth. Furthermore, three enrollment status categories were identified: dropout, academically at-risk (enrolled), and control. A self report survey was used to assess both level and intensity of volatile solvent use. Findings indicated that a higher proportion of the dropout cohort have used volatile solvents, used volatile solvents regularly, and used volatile solvents with more intensity than either the academically at-risk group or the control group. An interaction between gender and ethnicity was also revealed; American Indian females reported higher lifetime prevalence and thirty-day prevalence than did American Indian males, whereas for both the

---

Scott C. Bates is Research Assistant, Tri-Ethnic Center for Prevention Research, Department of Psychology, Colorado State University. Bradford Wong Plemons is Research Assistant, Department of Psychology, Western Washington University. Pamela Jumper-Thurman is Research Associate and Fred Beauvais is Senior Research Scientist, both with the Tri-Ethnic Center for Prevention Research, Colorado State University.

[Haworth co-indexing entry note]: "Volatile Solvent Use: Patterns by Gender and Ethnicity Among School Attenders and Dropouts." Bates, Scott C. et al. Co-published simultaneously in *Drugs & Society* (The Haworth Press, Inc.) Vol. 10, No. 1/2, 1997, pp. 61-78; and: *Sociocultural Perspectives on Volatile Solvent Use* (ed: Fred Beauvais, and Joseph E. Trimble) Harrington Park Press, an imprint of The Haworth Press, Inc., 1997, pp. 61-78. Single or multiple copies of this article are available for a fee from The Haworth Document Delivery Service [1-800-342-9678, 9:00 a.m. - 5:00 p.m. (EST). E-mail address: getinfo@haworth.com].

 

Mexican-American and White American samples, males report higher rates than females. Findings are discussed in terms of the influence of volatile solvent abuse and school success as well as previous findings.
*[Article copies available for a fee from The Haworth Document Delivery Service: 1-800-342-9678. E-mail address: getinfo@haworth.com]*

## *INTRODUCTION*

There has been a modest amount of research within the last decade focusing on the use and abuse of solvents (e.g., Crider & Rouse, 1988; Sharp, Beauvais, & Spence, 1992). Researchers have found a consistent relationship between volatile solvent abuse (see Beauvais and Oetting, 1987 for a note on terminology), poor performance in school (e.g., Creson & Welch cited in Creson, 1992; Epstein & Wieland, 1978; Fredlund, 1992; Korman, Matthews, & Lovitt, 1981; Oetting & Webb, 1992; Watson, 1977), truancy (e.g., Fredlund, 1992; Watson, 1977), and dropout rates (Creson & Welch cited in Creson, 1992; Epstein & Wieland, 1978; Fredlund, 1992; McSherry, 1988). The association of solvent inhalation with poor academic performance, truancy, and dropout rates has also been noted in several other countries (e.g., Annis & Watson, 1975; Chadwick, Yule, & Anderson, 1990; Wada & Fukui, 1993). The present study was designed to further explore the relationship between solvent use and school status within the U. S., with a special emphasis on American Indian (AI), Mexican-American (MA), and White American (WA) youth and gender patterns within these groups.

### *Dropout Rates*

McMillen et al. (1992) reported that national dropout rates have declined over the last 10 to 15 years. In 1992, about 11.0% of all persons ages 16 through 24 had not completed high school and were not enrolled in school. This continues a trend of decreasing rates; the rates in 1968 and 1979 were 16% and 14.6%, respectively. Reductions were noted across all ethnic groups although Rumberger (1987) cites data indicating recent increases in dropout for some minority groups.

The national dropout rates reported by McMillen et al. (1992) are much lower than the rates reported by the SCNAC (1986, 1987). In 1983, the SCNAC found that about one-fourth of U.S. students do not graduate from high school on schedule–the dropout rates varied from state to state and ranged between 5.2% and 42.8%.

Aside from historical trends, the dropout rate for Hispanics and American Indians remain considerably higher than for majority youth. Rates for

Hispanics have been reported at around 18% nationally (McMillen et al. 1992), with some localized rates as high as 60% (Colardarci, 1983). Rates specifically for Mexican Americans run as high as 45% (Chavez, Edwards, and Oetting, 1989). Steinberg, Blinde, and Chan (1984) indicate that the dropout rates for Mexican Americans are high regardless of socioeconomic status.

A review of 10 studies conducted since 1959 revealed that dropout rates among American Indian youths ranged between 15% and 60% (Office of Technology Assessment, 1990), while Chavers (1991) put the overall rate at 50%. Coladarci (1983) reported that in many American Indian schools, a dropout rate of 60% was not uncommon.

These data leave little doubt that school dropout is a serious problem among youth in the U. S., and that minority youth are especially at risk. It is also likely that school dropouts experience other social problems, including substance abuse.

### School Problems

Academic difficulty is often cited as a reason that youths drop out of school (Elliott, Voss, & Wendling, 1966; Gade, Hurlburt, & Fuqua, 1986; Kaplan & Luck, 1977; Morris, Ehren, & Lenz, 1991; Pittman, 1991). It is reasonable to assume that if students do not feel that they can succeed in school, they make the decision that school is not worth their time, and dropout. Kaplan and Luck (1977) reported that 50% of high school dropouts, at some time in their schooling, were held back a grade. Several authors (Morris, Ehren, & Lenz, 1991; Pittman, 1991) attempted to build models that would predict which students would drop out of high school. Examining the performance of students from 4th grade to 8th grade, Morris, Ehren, and Lenz (1991) discovered that, at all grade levels, getting poor grades significantly predicted dropping out. Ekstrom, Goertz, Pollack, and Rock (1986) summarized data collected by the National Center for Education Statistics and reported that dropouts had lower grades and tests scores, did less homework, and had more disciplinary problems than students who stayed in school. It would appear that the student who struggles in school, is held back, or has problems studying and/or testing, is more likely to leave school than the student who is well-adjusted academically. The similarity in profile between academically troubled youth and school dropouts suggests that these groups may be similar in other dimensions and in experiencing social problems.

### Solvent Abuse

A number of studies have compared solvent abuse rates between White American, American Indian, and Hispanic youth. Over thirty years ago,

Sterling (1964) reported that "glue sniffing" was predominantly an Anglo youth problem. More recently, Mason (1979) conducted an exploratory study of patterns of volatile substance abuse in six communities, and found that users tended to be young, male, and Mexican-American or White American. These findings are consistent with national lifetime prevalence rates for minority populations (see Beauvais, 1992a). Beauvais (1992b) reported that White Americans (17%) and American Indians (17%) report the highest incidences of lifetime prevalence use for solvents. Spanish-Americans (16%) were next in lifetime prevalence followed by Mexican-Americans (12%). The trends reported by Beauvais closely resemble those found by Frank et al. (1988). However, Mata and Andrew (1988) did not find significant differences between the rates of solvent use for Mexican-Americans (11%) and White Americans (9%).

Studies on gender and volatile solvent use are sparse, but use among females does appear to be increasing. This trend may be particularly significant because of potential consequences related to health and pregnancy. Beauvais (1992a) noted that the rates of solvent use for males and females have been converging over the past 20 years. Cohen, in 1973, estimated that 10 times more males than females used solvents. But in 1988, the National High School Senior Survey reported lifetime prevalence rates of 19.5% for males and 14% for females (Johnston et al. 1989). Likewise, the National Household Survey of 12- to 17-year-olds, conducted in 1988, reported volatile solvent use rates of 9.2% for males and 8.3% for females (National Household Survey, 1990). Edwards and Oetting (1995) reported a greater discrepancy in lifetime prevalence between males (15.1%) and females (8.5%), although the overall rates of use were similar to other studies.

In a study involving 7th-12th grade American Indian adolescents, lifetime prevalence rates for males were reported as 31.8%, and for females, 32.9% (Beauvais, Oetting, & Edwards, 1985). Of particular interest was the 30-day prevalence rates, 9.1% for females and 7.6% for males. In yet another study examining trends in Indian adolescent drug and alcohol use, both reservation and non-reservation, Beauvais (1992b) reported slightly higher use among reservation twelfth grade females as compared to males (21% and 20% respectively), and equal or higher use among eighth grade females, both reservation (35% and 35%) and non-reservation (22% and 18% respectively) as compared to their male counterparts.

### *Solvent Abuse Related Problems*

Stephens, Diamond, Spielman, and Lipton (1978) reported that there was a strong consistent relationship between low grade point average and solvent use. At almost all grade levels solvent users are mostly "D" or "F" students. These findings are consistent with results found by others (e.g., Coulehan et al. 1983; Frank, Marel, & Schmeidler, 1988). Coulehan et al. (1983) reported that Navajo, high school youths were more likely to engage in gasoline sniffing if they had failed courses. Wingert and Fifield (1985) compared the grade point averages of American Indian non-users, one-time users, and repeat users of volatile solvents and found that non-users attained higher grade point averages than one-time users and repeat users–surprisingly, there was no significant differences between grade point averages of one-time and repeat users. Wingert and Fifield concluded that "the finding of significant differences among the groups on the grade point average should be interpreted not only as a logical conclusion based upon the differences in the academic achievement areas, but may also more broadly represent factors other than academic skills" (p. 1580).

Other characteristics of volatile solvent abusers–related to poor academic performance–are apparent. Wada and Fukui (1993) found that more lifetime users than non-users of volatile substances believe that their school life is "rather unpleasurable" or "completely unpleasurable," and that non-users were more likely to participate in school activities than lifetime users. Students who do not expect to graduate are more likely to have used solvents and have done so in the last 6 months when compared with students who do expect to graduate. Stephens et al. (1978) found that the rate of solvent use drops as a students' education level rises; they also found that those who had not completed primary and secondary education had the highest use rates. Among primary school boys, Nurcombe, Bianchi, Money, and Cawte (1970) reported that users of volatile solvents tended to be slower learners than non-users of volatile solvents, and that non-users were rated average or above average in scholastic progress more so than users; yet, these findings were not statistically significant, possibly due to both the small sample size of 22 matched subjects and the use of a non-directional rather than a directional test. Winburn and Hays (1974) found a significant correlation between the grade level completed when a student drops out of school and past use of drugs–the earlier an individual drops out of school, the more likely he/she is to have used drugs.

Simpson (1992) followed a high-risk group of 175 Mexican-American youths entering an advocacy program. Sixty-three percent of the youths admitted using volatile solvents. It was also found that 71 percent of the high-risk sample had dropped out of school, only 9 percent had graduated or received their GED, and 15 percent were still enrolled in school or a formal vocational training program. Simpson, however, did not compare the dropout rates of solvent users with non-users. Menon, Barrett, and Simpson (1990) found that problems in school performance and school conduct were two of the variables that played a significant role in predicting volatile substance abuse among 599 Mexican-American youth admitted to drug abuse programs in Texas. Menon et al. also showed how volatile substance users skipped classes and were sent home for breaking school rules significantly more often than non-users.

A trend for higher use of solvents in areas of lower socioeconomic status was found for Mexican-American youth (Bachrach & Sandler, 1985). Padilla et al. (1979) compared the volatile solvent use rates of 457 Mexican-American youth in the barrio to volatile solvent use rates from a national sample–they found that volatile solvent users in barrios were more likely to have ever used volatile solvents, and are at least 14 times more likely to be currently using volatile solvents. Hispanics in poor barrio environments may use solvents heavily, but Hispanics in less stressful environments do not. Thus, the high use rates of Hispanic youths living in barrios may be due to the socioeconomic conditions to a greater extent than school attendance or completion.

Comparing a sample of high school dropouts with a random sample of 10th, 11th, and 12th graders, Winburn and Hays (1974) found that dropouts (18.2%) had tried solvents to a greater extent than any of the three cohorts (13.2%, 11.9%, and 11.5%, respectively). Winburn and Hays also found dropouts had higher six month prevalence use rates than the 10th through 12th grade cohorts. The SCNAC (1987) describe how "dropouts and absentees tend to use drugs more frequently and they appear more often in treatment centers than other high school students. Thus, it is easy to establish a correlation between drugs and dropping out . . . What the nature of this relationship is, which is the cause and which is the effect, is less evident" (p. 11). Clearly, further research is necessary to establish a causal relationship between variables of school status and volatile solvent abuse among minority groups and across gender.

## *METHOD*

### *Participants*

The current study is an attempt to further understand the relationship between solvent/inhalant use among White American (WA) and two ethnic minority groups in the western United States: American Indians (AI) and Mexican-Americans (MA). Data were collected from 1988 to 1992, via administration of self-report surveys and examination of school records. For the American Indian cohort, data were collected in 6 different sites in the Western United States, four sites were on reservations, one was a high school in a mid-sized city with a high percentage of American Indian students, and the sixth location was a rural area with a large American Indian population. The Mexican-American and White American samples were collected at 3 sites: a large metropolitan city, a mid-sized town, and a small rural town–all were located in the western United States.

Three sub-populations of participants were identified within each group: high school dropouts (DO), academically at-risk non-dropouts (AR), and non-dropout controls (CO) students. Dropout respondents were defined as youths who were absent from school–without contact–for one month or more. Each dropout subject was matched to a student of similar academic background but who had not dropped out of school (academically "at-risk"), and a control student who was enrolled in school and showing no academic problems. Participants in the academically at-risk group were selected to match the dropout sample in terms of grade point average (as closely as possible), ethnicity, sex, grade in school, and age, but were still enrolled in school. This represented an attempt to match youth with similar academic difficulties with status as a dropout or non-dropout. Control participants were selected to match a DO respondent by ethnicity, grade in school (last grade attended by dropout) and sex. The participant pool (N = 3416) was 53.4% male (n = 1825) and 46.6% (n = 1591) female. Sample sizes for all cohorts are presented in Table 1. Participants ranged in age from 12 to 21 years with an overall mean age of 16.56 (*SD* = 1.33); see Table 2 for breakdown of age by group. The mean GPA for each cohort is presented in Table 3.

### *Instruments*

The data for this study came from an ongoing dropout project conducted by the Tri-Ethnic Center for Prevention Research at Colorado State University. In that project an extensive, multiscale instrument was administered to participants and school records were examined for academic information. This study utilized the items asking about involvement with solvents, including lifetime and 30 day prevalence, age of first use, style of use and questions about psychosocial factors that may be related to solvent abuse.

### *Procedure*

Dropout respondents either came to a school or were met in another public building to complete the survey; data were collected from comparison respondents (AR and CO) during school hours. The surveys were completed anonymously and took about 90 minutes to finish. Upon completion of the survey, and in the presence of the field researcher, the survey was sealed in an envelope and immediately mailed to the Tri-Ethnic Center for data entry and processing. Dropout participants were paid $20 for participation, control and academically at-risk participants were paid $10.

TABLE 1. Sample Size

| | Mexican-American | | White American | | American Indian | |
|---|---|---|---|---|---|---|
| Enrollment Status | Male | Female | Male | Female | Male | Female |
| Control | 301 | 227 | 125 | 129 | 110 | 133 |
| At risk | 350 | 236 | 118 | 113 | 120 | 135 |
| Dropout | 380 | 268 | 171 | 148 | 113 | 158 |

TABLE 2. Mean Age of All Groups

| | Mexican-American | | White American | | American Indian | |
|---|---|---|---|---|---|---|
| Enrollment Status | Male | Female | Male | Female | Male | Female |
| Control | 16.49 | 16.52 | 16.77 | 16.70 | 16.11 | 15.93 |
| At risk | 16.46 | 16.38 | 16.82 | 16.59 | 16.40 | 16.16 |
| Dropout | 16.67 | 16.49 | 16.85 | 16.64 | 17.41 | 16.99 |

TABLE 3. Mean Grade Point Average of All Groups

| | Mexican-American | | White American | | American Indian | |
|---|---|---|---|---|---|---|
| Enrollment Status | Male | Female | Male | Female | Male | Female |
| Control | 2.48 | 2.53 | 2.49 | 2.58 | 2.36 | 2.59 |
| At risk | 1.28 | 1.51 | 1.43 | 1.77 | 1.37 | 1.55 |
| Dropout | .87 | 1.07 | 1.17 | 1.38 | 1.14 | 1.36 |

Note: For 481 of the respondents, 14.1% of the sample, no GPA was reported.

Participants were assured of confidentially and were asked to sign a consent form; parent permission was obtained for respondents under the age of 18.

## RESULTS

### Prevalence

Table 4 shows lifetime prevalence rates for volatile solvent use by ethnicity, gender, and enrollment status. Table 5 shows the significant differences in lifetime prevalence as determined by odds ratios and confidence intervals (CI) for each odds ratios derived from logistic regression. If the CI (95%) does not include an odds ratio of 1.00 (a 1:1 relationship), we can assume that a significant difference exists. Significant differences in lifetime prevalence were found for both sex and ethnicity. These results, however, were accounted for by the interaction discussed below.

Across both gender and ethnicity there were significant differences for enrollment status ($p < .05$). High school dropouts were 2.27 times more likely than control respondents to have used volatile solvents in their lifetime, academically at-risk respondents were 1.72 times more likely than control subjects to have used volatile solvents in their lifetime, and dropout participants were 1.32 times more likely than academically at-risk respondents to have used volatile solvents in their lifetime.

The interaction between gender and ethnicity revealed by the analysis (see Table 5) is shown graphically in Figure 1. For both MA and WA

TABLE 4. Percentage of Respondents Indicating Lifetime Prevalence for Volatile Solvent Use

| | DO | AR | CO |
|---|---|---|---|
| Mexican-American Male | 37.3 | 29.8 | 19.7 |
| Mexican-American Female | 28.1 | 22.3 | 14.8 |
| White American Male | 46.1 | 32.8 | 28.8 |
| White American Female | 28.6 | 31.3 | 18.8 |
| American Indian Male | 34.5 | 32.0 | 11.7 |
| American Indian Female | 40.2 | 32.6 | 23.9 |

Note: DO = dropout, AR = academically at-risk, CO = control.

TABLE 5. Odds Ratios and 95% Confidence Intervals for Lifetime Prevalence of Volatile Solvent Use

| Predictor | Odds Ratio | CI (95%) |
|---|---|---|
| Sex (F:M) | 1.18 | 1.00-1.38 |
| Ethnicity (MA:WA) | 1.33 | 1.11-1.45 |
| Enrollment Status | | |
| Dropout:Control | 2.27 | 1.86-2.76 |
| Academically At Risk:Control | 1.72 | 1.40-2.11 |
| Dropout:Academically At Risk | 1.32 | 1.09-1.57 |
| Ethnicity by Sex | | |
| Mexican-American (M/F):American Indian (M/F) | 1.43 | 1.18-1.74 |
| White American (M/F):American Indian (M/F) | 1.51 | 1.21-1.89 |

Note: All other main effects and interactions were non-significant.

respondents, a higher proportion of males than females reported lifetime use, however, for AI respondents this difference was reversed: for the AI cohort sampled, a higher proportion of females reported use of volatile solvents at least once in their lifetime.

Thirty day prevalence is an indication of recent volatile solvent use and includes those youth who are more chronically involved. Table 6 shows 30-day prevalence rates for volatile solvent use by ethnicity, sex and enrollment status. Logistic regression demonstrated several differences among groups (see Table 7). Across the three ethnic groups dropout participants and academically at-risk participants show significantly ($p < .05$) higher thirty-day prevalence rates for volatile solvent use than control students; in each case at least twice the proportion of DO and AR respondents, as compared to CO respondents, reported using in the last month. Thus, high school dropouts were 2.88 times more likely than control respondents to have used volatile solvents in the last month, and academically at-risk respondents were 2.21 times more likely than control subjects to have used volatile solvents in the last month. No difference in recent use was detected between DO and AR cohorts.

An interaction between gender and ethnicity was revealed by the analysis. Figure 1 graphically represents thirty-day prevalence rates by gender and ethnicity (across enrollment status). For MA respondents, a higher

FIGURE 1

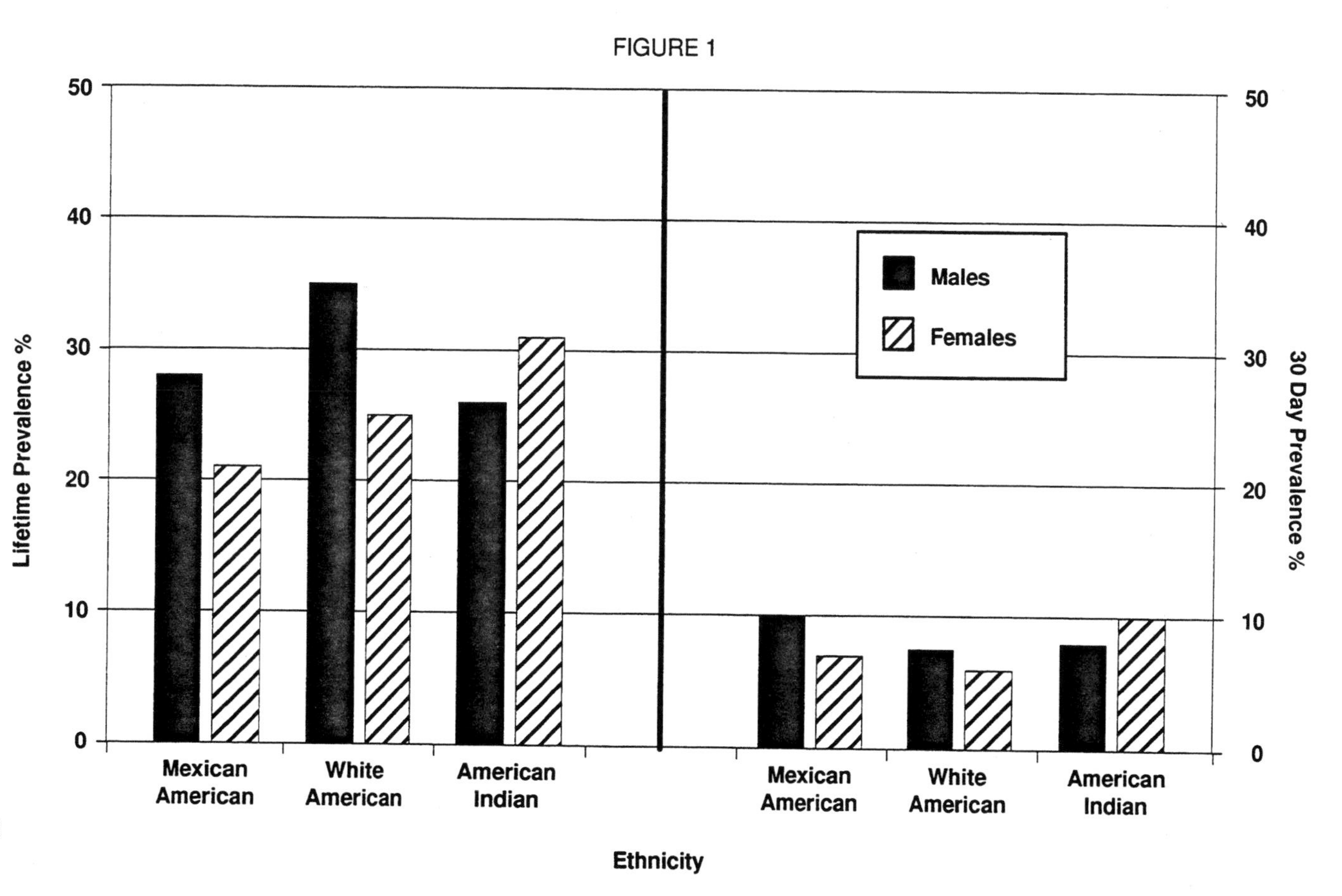

Lifetime Prevalence %
30 Day Prevalence %
50
40
30
20
10
0
Males
Females
Mexican American
White American
American Indian
Ethnicity

TABLE 6. Percentage of Respondents Indicating Thirty-Day Prevalence for Volatile Solvent Use

| | DO | AR | CO |
|---|---|---|---|
| Mexican-American Male | 16.6 | 9.0 | 5.3 |
| Mexican-American Female | 9.5 | 8.2 | 3.1 |
| White American Male | 12.1 | 6.0 | 4.8 |
| White American Female | 6.8 | 6.3 | 3.1 |
| American Indian Male | 9.1 | 11.8 | 3.7 |
| American Indian Female | 10.8 | 15.0 | 5.3 |

Note: DO = dropout, AR = academically at risk, CO = control.

TABLE 7. Odds Ratios and 95% Confidence Intervals for Thirty-Day Prevalence of Volatile Solvent Use

| Predictor | Odds Ratio | CI (95%) |
|---|---|---|
| Enrollment Status | | |
| Dropout:Control | 2.88 | 2.03-4.09 |
| At Risk:Control | 2.21 | 1.53-3.19 |
| Ethnicity by Sex | | |
| Mexican-American (M/F):American Indian (M/F) | 1.42 | 1.05-1.93 |

Note: All other main effects and interactions were non-significant.

proportion of males than females reported use in the month before the survey was completed, however, for AI respondents, this difference was reversed. Thus, as was the case with lifetime prevalence, AI females reported more recent volatile solvent use than AI males. No other significant main effects or interactions were found.

### *Patterns of Use*

Several additional variables were analyzed in an effort to understand the interactions found in both lifetime and 30-day prevalence; anxiety, alienation, anger, blame, depression, deviant behavior, school adjustment, self-esteem, and tolerance of deviance, were all included in the analyses.

None, however, were found useful with respect to understanding the interactions presented above. That is, the variables found to be associated with solvent use in the literature did not predict the differences found between males and females across ethnic groups.

Patterns of volatile solvent use were investigated with the use of items targeting age of first use and intensity of use. Only respondents indicating that they had "huffed" or "sniffed" at least once in their lives were selected for these analyses, resulting in 954 cases (28.1% of the total sample). There were no significant differences in mean age of first use by gender (male = 13.1, female = 13.2), ethnic group (MA = 13.3, WA = 13.1, AI = 13.0), or enrollment status (DO = 13.3, AR = 13.0, CO = 13.1).

Intensity of solvent use was measured with the question, "When you sniff to get high, how do you like to do it?" Responses were scaled from 1 (I never do it) to 5 (to stay high for several hours at a time); those who had tried using volatile solvents generally reported that they use "just enough to feel it a little bit." However, while the effect was small, a two-way ANOVA revealed a main effect for enrollment status ($F(2, 919) = 7.36, p < .001$, $Eta2 = .02$). Subsequent Tukey HSD post-hoc tests ($p < .05$) revealed that DO participants ($M = 2.15$, $SD = 1.09$) reported using volatile solvents more intensely than either AR ($M = 2.07$, $SD = 1.00$) or CO ($M = 1.82$, $SD = .90$) respondents. Likewise, AR respondents indicated more intense volatile solvent use than CO respondents. However, there were no differences in intensity of use by gender.

## *DISCUSSION*

The greater lifetime prevalence and thirty-day prevalence rates for dropouts and controls are consistent with findings by other researchers (e.g., Bachrach & Sandler, 1985; Medina-Mora et al., 1978; SCNAC, 1987; Winburn & Hays, 1974). Indeed, it is apparent that a relationship exists between enrollment status and volatile solvent abuse–youths who drop out of school are more likely to have ever used volatile solvents than their in-school counterparts, and thirty-day prevalence rates show that dropouts are more likely to be currently involved with volatile solvents than controls. While this is a confirmatory finding, it is nevertheless important since it firmly establishes this relationship across three ethnic groups using the same methodology.

Several studies have indicated that trouble with school is associated with volatile solvent use (e.g., Coulehan et al. 1983; Stephens et al. 1978; Wingert & Fifield, 1985). Thus, solvent use rates of at-risk youths across the sample should be higher than those of control youth. Indeed, at-risk youths were found to be 1.72 times more likely than controls to have used volatile solvents during their lifetime, and were 2.21 times more likely than controls to have used volatile solvents during the last 30 days. Higher use rates for at-risk youth, compared with control youth, should not be solely attributed to lower grade point averages because a causal relationship has not been established between the volatile solvent use rates of the two academic status categories. Dropouts reported higher lifetime prevalence rates of volatile solvents compared to their at-risk youth counterparts. This finding indicates that being academically at risk may not be entirely predictive of tendencies toward deviance, such as volatile solvent use. It is quite possible that some youth receive low grades due to problems such as illness or learning disabilities.

The interaction between ethnicity and gender, indicating that American Indian females have higher rates of volatile solvent use than males, is an important finding because it counters what was found in other groups (although May and Del Vecchio, in this volume, found convergence across gender for their Mexican-American sample). The literature, to date, has generally documented higher rates among males, although the gap has been closing in recent years (Beauvais, 1992a). Exceptions, however, were reported during the last decade, where rates for Indian females were found to be essentially identical to that for males (Beauvais, Oetting & Edwards, 1985, Beauvais, 1992a, 1992b). The present results indicate a long term trend wherein Indian females have surpassed Indian males in their use of volatile solvents. Other variables, including several social and mental health measures, were examined to explain this finding, but no additional explanation for this interaction was derived. Oetting and Webb (1992) found that solvent users, in general, have been described as having higher rates of dysfunction in other areas, and it may be that American Indian females are increasingly being subjected to a wider array of environmental stressors, resulting in higher use becoming more acceptable for females in this population. It may also be suggested that American Indian females, who are involved in deviant behaviors, are more prone to involve volatile solvent use in that lifestyle, thus raising the potential of fetal effects and other health risks. Regardless, the complexity of this issue warrants exploration of a wide range of potential factors to help explain this finding. Specific issues are: How might American Indian female profiles and use patterns differ from males? Are their peer dynamics different? Are changing cultural roles involved? What is the meaning of volatile solvent use among Indian females? Most risk factor studies have utilized male samples or reported results using combined samples of males and females–

perhaps an in-depth risk factor study should be completed exclusively on American Indian females.

Future studies, of a longitudinal nature, although expensive in terms of time and effort, would certainly be useful in clarifying the results found here. It would also be useful to know the causative link between substance abuse, poor academic performance, dropping out and other forms of deviance. Furthermore, a longitudinal study of Indian youth would also help shed light on the disparity in rates found in this study. General population surveillance must also be maintained to determine whether the convergence between males and females in solvent use is restricted to American Indian youth or whether it may also be occurring among other youth.

Practitioners must certainly be alert to the fact that those experiencing problems with school are more likely to be harmfully involved with volatile solvents. Additionally, when an enduring pattern of solvent abuse is detected among youth, there is a strong likelihood that there is a co-occurrence of a wide range of other social and psychological problems.

## REFERENCES


Annis, H. M., & Watson, C. (1975). Drug use and school dropouts: A longitudinal study. *Canadian Counseling, 9*, 155-162.

Bachrach, K. M., & Sandler, I. N. (1985). A retrospective assessment of volatile solvent abuse in the Barrio: Implications for prevention. *The International Journal of the Addictions, 20(8)*, 1177-1189.

Beauvais, F. (1992). Volatile solvent abuse: Trends and patterns. In C. W. Sharp, F. Beauvais, & R. Spence (Eds.), *Inhalant Abuse: A Volatile Research Agenda.* Rockville, MD: National Institute on Drug Abuse Research, (NIH #93-3475).

Beauvais, F. (1986). Social and psychological characteristics of inhalant abusers. Paper presented at World Health Organization Advisory Group meeting. "Adverse Health Consequences of Volatile Solvents/Inhalants," Mexico City.

Beauvais, F., and Oetting, E. (1987). Toward a clear definition of inhalant abuse. *The International Journal of the Addictions, 22,* 779-784.

Brill, N. W., and Christie (1974). Marijuana use and psychosocial adaptation. *Archives of General Psychiatry,* 31, 713-719.

Bruno, J. E., & Doscher, L. (1979). Patterns of drug use among Mexican-American potential school dropouts. *Journal of Drug Education,* 9(1), 1-10.

Chadwick, O., Yule, W., & Anderson, R. (1990). The examination attainments of secondary school pupils who abuse solvents. *British Journal of Educational Psychology, 60,* 180-191.

Chavers, D. (1991). Indian education: Dealing with a disaster. *Principal, 70,* 28-29.

Chavez, E. L., Edwards, R. W., and Oetting, E. R. (1989). Mexican-American and white American dropouts' drug use, health status and involvement in violence. *Public Health Rep, 104,* 594-604.

Cohen, S. (1973). The volatile solvents. *Public Health Review, 2*, 185-214.

Coulehan, J. L., Hirsch, W., Brillman, J., Sanandria, J., Welty, T. J., Colaiaco, P., Koros, A., & Lober, A. (1983). Gasoline sniffing and lead toxicity in Navajo adolescents. *Pediatrics, 71(7)*, 113-117.

Creson, D. L. (1992). Comments on psychosocial characteristics. In C. W. Sharp, F. Beauvais, & R. Spence (Eds.), *Inhalant Abuse: A Volatile Research Agenda.* Rockville, MD: National Institute on Drug Abuse Research, (NIH #93-3475).

Crider, R. A., & Rouse, B. A. (Eds.). (1988). *Epidemiology of Inhalant Abuse: An Update.* Rockville, MD: National Institute on Drug Abuse Research.

Edwards, R. & Oetting, E. (1995). Inhalant use in the United States. In Kozel, N., Sloboda, Z., & De La Rosa, M., (Eds.), *Epidemiology of Inhalant Abuse: An International perspective.* Rockville, MD: National Institute on Drug Abuse Research, (NIH #95-3831).

Epstein, M. H., & Wieland, W. F. (1978). Prevalence survey of inhalant abuse. *The International Journal of the Addictions, 13(2)*, 271-284.

Frank, B., Marel, R., & Schmeidler, J. (1988). The continuing problem of youthful solvent abuse in New York state. In R. A. Crider, & B. A. Rouse (Eds.), *Epidemiology of Inhalant Abuse: An Update.* Rockville, MD: National Institute on Drug Abuse Research.

Frase, M. J. (1989). Dropout rates in the United States: 1988 (NCES 89-609). Washington: National Center for Education Statistics.

Fredlund, E. V. (1992). Epidemiology of volatile solvent abuse: The Texas experience. In C. W. Sharp, F. Beauvais, & R. Spence (Eds.), *Inhalant Abuse: A Volatile Research Agenda.* Rockville, MD: National Institute on Drug Abuse Research, (NIH #93-3475).

Johnston, L. D., O'Malley, P. M., & Bachman, J. G. (1994). National survey results on drugs use from the monitoring the future study, 1975-1993. Rockville, MD: National Institute on Drug Abuse. (NIH# 94-3809).

Johnston, L., O'Malley, P., & Bachman, J. (1989). Drug use, drinking and smoking: National survey results from high school, college, and young adult populations, 1975-1988. Rockville, MD: National Institute on Drug Abuse. (ADM # 89-1638).

Kandel, D. B. (1978). Convergences in prospective longitudinal surveys of drug use in normal populations. In D. B. Kandel (Ed.), *Longitudinal Research on Drug Use.* Washington, DC: Hemisphere Publishing Corporation.

Kaufman, P., & Frase, M. J. (1990). Dropout rates in the United States: 1989 (NCES 90-659). Washington: National Center for Education Statistics.

Korman, M., Matthews, R. W., & Lovitt, R. (1981). Neuropsychological effects of abuse of inhalants. *Perceptual and Motor Skills, 53*, 547-553.

Mason, T. (1979). Inhalant use and treatment. Rockville, MD: National Institute on Drug Abuse.

Mata, A. G., & Andrew, S. R. (1988). Inhalant abuse in a small rural south Texas community: A social epidemiological overview. In R. A. Crider, & B. A. Rouse (Eds.), *Epidemiology of Inhalant Abuse: An Update.* Rockville, MD: National Institute on Drug Abuse Research.

McMillen, M. M., Kaufman, P., Hausken, E. G., & Bradby, D. (1993). Dropout rates in the United States: 1992 (NCES 9-3-464). Washington: National Center for Education Statistics.

McSherry, T. M. (1988). Program experiences with the solvent abuser in Philadelphia. In R. A. Crider, & B. A. Rouse (Eds.), *Epidemiology of Inhalant Abuse: An Update.* Rockville, MD: National Institute on Drug Abuse Research.

Medina-Mora, M. E. I., Schnaas, L. A., Terroba, G. T., Isoard, Y. V., & Suarez, C. U. (1978). Epidemiology of inhalant use in Mexico. In C. W. Sharp & L. T. Carroll (Eds.), *Voluntary Inhalation of Industrial Solvents.* Rockville, MD: National Institute on Drug Abuse Research.

Menon, R., Barrett, M. E., & Simpson, D. D. (1990). School, peer group, and inhalant use among Mexican American adolescents. *Hispanic Journal of Behavioral Sciences, 12* (*4*), 408-421.

National Institute on Drug Abuse, (1990). National household survey on drug abuse: Main findings 1988. Rockville, MD: National Institute on Drug Abuse. (ADM #90-1682).

Nurcombe, B., Bianchi, G., Money, J., & Cawte, R. (1970). A hunger for stimuli: The psychosocial background of petrol inhalation. *British Journal of Medical Psychology and Psychiatry,* 43, 367-374.

Oetting, E. R., & Beauvais, F. (1987a). Common elements in youth drug abuse: Peer clusters and other psychosocial factors. *Journal of Drug Issues, 17(1 & 2),* 133-151.

Oetting, E. R., & Beauvais, F. (1987b). Peer cluster theory, socialization characteristics and adolescent drug use: A path analysis. *Journal of Counseling Psychology, 34*(*2*), 205-213.

Oetting, E. R., & Beauvais, F. (1988). Common elements in youth drug abuse. In S. Peele (Ed.), *Visions of Addiction* (pp. 141-161). Lexington: D.C. Heath and Company.

Oetting, E. R., & Webb, J. (1992). Psychosocial characteristics and their links with inhalants: A research agenda. In C. W. Sharp, F. Beauvais, & R. Spence (Eds.), *Inhalant Abuse: A Volatile Research Agenda.* Rockville, MD: National Institute on Drug Abuse Research, (NIH #93-3475).

Oetting, E. R., & Webb, J. (1992). Psychosocial characteristics and their links with inhalant use: A research agenda. In C. Sharp, F. Beauvais, & R. Spence (Eds.), *Inhalant Abuse: A Volatile Research Agenda* (pp. 59-98). Washington, D.C.: National Institute on Drug Abuse, (NIH #93-3475).

Padilla, E. R., Padilla, A. M., Morales, A., Olmedo, E. L., & Ramirez, R. (1979). Inhalant, marijuana, and alcohol abuse among barrio children and adolescents. *International Journal of the Addictions, 14* (*7*), 945-964.

Select Committee on Narcotics Abuse and Control (1986). Drugs and dropouts (SCNAC-99-2-2). Washington, DC: U.S. Government Printing Office.

Select Committee on Narcotics Abuse and Control (1987). 1987 update on drugs and dropouts (SCNAC-100-1-17). Washington, DC: U.S. Government Printing Office.

Sharp, C. W., Beauvais, F., & Spence, R. (Eds.). (1992). *Inhalant Abuse: A Volatile Agenda.* Rockville, MD: National Institute on Drug Abuse Research.

Simpson, D. D. (1992). A longitudinal study of inhalant use: Implications for treatment and prevention. In C. W. Sharp, F. Beauvais, & R. Spence (Eds.), *Inhalant Abuse: A Volatile Research Agenda.* Rockville, MD: National Institute on Drug Abuse Research, (NIH #93-3475).

Stephens, R. C., Diamond, S. C., Spielman, C. R., & Lipton, D. S. (1978). Sniffing from Suffolk to Syracuse: A report of youthful solvent use in New York state. In C. W. Sharp & L. T. Carroll (Eds.), *Voluntary Inhalation of Industrial Solvents.* Rockville, MD: National Institute on Drug Abuse Research.

Sterling, J. W. (1964). A comparative examination of two modes of intoxication–an exploratory study of glue sniffing. *Journal of Criminal Law, Criminology, & Police Science, 55,* 94-99.

Trimble, J. E. (1995). Toward an understanding of ethnicity and ethnic identity, and their relationship with drug use research. In G. J. Botvin, S. Schinke, & M. A. Orlandi (Eds.), *Drug Abuse Prevention with Multiethnic Youth.* Thousand Oaks, CA: Sage Publications.

Wada, K, & Fukui, S. (1993). Research report: Prevalence of volatile solvent inhalation among junior high school students in Japan and background life style of users. *Addiction, 88,* 89-100.

Watson, J. M. (1977). 'Glue sniffing' in profile. *The Practitioner, 218,* 255-259.

Winburn, G. M., & Hays, J. R. (1974). Dropouts: A study of drug use. *Journal of Drug Education, 4 (2),* 249-254.

Wingert, J. L., & Fifield, M. G. (1985). Characteristics of Native American users of inhalants. *The International Journal of the Addictions, 20(10),* 1575-1582.

# The Inhalant Dilemma: A Theoretical Perspective

Bernard Segal, PhD

**SUMMARY.** A comprehensive model describing drug abuse behavior is presented which incorporates biological, social and psychological domains. Within the structure of this interactive model, specific drug related phenomena are discussed, including inhalant abuse, cultural factors (with an emphasis on Alaskan Natives) and post-traumatic stress disorder (PTSD). Future research needs are addressed with the presentation of several hypotheses derived from the model. *[Article copies available for a fee from The Haworth Document Delivery Service: 1-800-342-9678. E-mail address: getinfo@haworth.com]*

Inhalants, or volatile solvents, are chemicals whose vapors, when inhaled, produce psychoactive effects. Many of these chemicals, such as spray paints, gasoline, paint thinner, and lighter fluid are not drugs in the ordinary sense of the term, but are industrial chemicals (solvents or aerosols) that are used as a drug because of their euphoric or intoxicating properties. Other products used are lacquer and lacquer thinner, acetone, hair sprays, spray starches, carbon tetrachloride and aerosol products. Many of these aerosols not only contain a conventional solvent, but also one of the freons–a chlorinated, fluorinated methane (Chlorofluorocar-

---

Bernard Segal is Director, Center for Alcohol and Addiction Studies, University of Alaska Anchorage, AK.

Author note: Parts of the theoretical model espoused herein are in large part an outgrowth of discussions with Ira H. Cisisn, PhD, held some years ago. I am indebted to him for his past helpfulness, and dedicate this paper to his memory.

[Haworth co-indexing entry note]: "The Inhalant Dilemma: A Theoretical Perspective." Segal, Bernard. Co-published simultaneously in *Drugs & Society* (The Haworth Press, Inc.) Vol. 10, No. 1/2, 1997, pp. 79-102; and: *Sociocultural Perspectives on Volatile Solvent Use* (ed: Fred Beauvais, and Joseph E. Trimble) Harrington Park Press, an imprint of The Haworth Press, Inc., 1997, pp. 79-102. Single or multiple copies of this article are available for a fee from The Haworth Document Delivery Service [1-800-342-9678, 9:00 a.m. - 5:00 p.m. (EST). E-mail address: getinfo@haworth.com].

 

bons [CFCs]) or ethane derivative–as a propellant. The medicinal or pharmacological chemicals often included in this category are anesthetics, such as nitrous oxide, and other compounds that are inhalable. Inhalants can be grouped into three categories: organic solvents (hydrocarbons), volatile nitrates, such as amyl and butyl nitrites, and nitrous oxide.

The most common methods of using these substances are: (a) placing the solvent in a plastic bag and inhaling the fumes ("bagging"), which is a very hazardous procedure because the user risks suffocation, and many deaths have been attributed to this method; (b) soaking a rag or handkerchief in the substance and inhaling the fumes; (c) sniffing or inhaling the fumes or vapors directly from the container; and (d) spraying the substance, such as a lacquer, into a soda pop can and inhaling the vapors directly ("huffing").

The general effect of these chemicals is a rapid, general depression of the CNS, characterized by marked inebriation, dizziness, floating sensations, exhilaration, and intense feelings of well-being. A breakdown in inhibitions and feelings of greatly increased power and aggressiveness can also take place; these sensations are very similar to alcohol intoxication. Vivid visual hallucinations may also be present. These effects last from about 15 minutes to a few hours, depending on the substance, the dose and duration of use. Chemical solvents can be highly toxic, and death or severe physical damage can result from their use. Sudden death can occur from heart or lung failure, asphyxiation, paralysis of breathing mechanisms, or accidents as a result of being inebriated. Prolonged use of these substances can lead to kidney, liver, lung and brain damage. There is no evidence that tolerance develops, and physical dependence has not been demonstrated.

The use of these solvents or inhalants has been predominantly associated with youth (ages 6 to 14), and the problem has emerged as one of national concern (Caputo, 1993), but it has been with us a long time (cf., Brecher, 1972). Inhalant abuse has also been an international problem, with reports of youth using inhalant substances in Japan, Russia, Europe, England, Scandinavia, Canada, Australia and in South American countries (Caputo, 1993; Segal, 1988). Inhalant use has largely been associated with males, but the gender gap is closing rapidly (Beauvais, 1992).

In the United States, Johnston, O'Malley and Bachman's 1992 National Survey of Secondary School Students (Johnston, O'Malley, & Bachman, 1993) reported an increase in inhalant use involving such substances as butane, solvents, glues, and nitrous oxide. They reported that "one in every six eighth-graders (17.4%) used some inhalant in their lifetime; and one in every 20 (4.7%) used one in the past 30 days" (p. 4).

A year later these researchers (University, 1994) continued to be concerned about increasing inhalant use among American youth, particularly those in lower grades. They stated, with respect to the increase in inhalant use in their most recent (1993) findings, that: "We don't think that young people fully understand the dangers of inhalants, perhaps because most of the substances inhaled are common household products, but they can definitely be lethal" (University, 1994, p. 3).

It is important to note that while drug-taking behavior has become a problem of national concern, particularly among youth, a characteristic of such behavior is that it is more prevalent among specific ethnic groups (cf. La Rosa & Adrados, 1993). American Indian and Alaskan Native youth, for example, show exceptionally high levels of drug use, principally alcohol and inhalants (Beauvais, 1992; Beauvais & Segal, 1992), while African Americans and Latino adults, who are also over represented among drug-using populations, evidence higher levels of narcotics and cocaine use.

## *A THEORETICAL PERSPECTIVE*

The questions that arise concerning youthful inhalant use are: What attracts youth to inhalants? Why do youth put themselves at risk by using inhalant substances? Why are some youth, specifically American Indian and Alaskan Native youth, more vulnerable to using inhalants than youth from other racial groups?

Presently, there are insufficient explanations of possible motivation and causation for the deliberate inhalation of volatile substances (Caputo, 1993). Nathan (1990) described the lack of efficacy of psychological theories of drug-taking behavior as follows: "valid predictors of onset and course of substance abuse have not been identified. . . . Performance on personality assessment instruments, projective and intelligence tests, and measures of neuropsychological functioning, as well as on life events and personal history measures, have not provided information on differences between persons who will develop substance abuse and those who will not" (p. 369). Inhalant use is no exception to this situation, and presents an even more complicated picture than the use of other drugs (Oetting & Webb, 1992).

Biological research has been equally limited in its ability to develop functional theories of drug-taking behavior. Despite advances in alcohol research, such as an understanding of the relationship between alcohol and aldehyde dehydrogenase genotypes and alcohol metabolism (cf., Segal et al. 1993), reliable functional differences in biological systems between persons who will develop alcoholism and those who will not have not been adequately described (Agarwal & Goedde, 1989; Nathan, 1990). Thus, no fundamental advances have been made regarding either a biological or behavioral basis of predicting drug users from nonusers, much less being able to identify youth who are potential inhalant users.

Perhaps part of the failure to achieve reliable biological or behavioral predictors is related to the fact that researchers studying drug-taking behavior largely conduct independent studies in which different disciplines pursue an explanation of the problem autonomously. Behavioral researchers, for example, tend to focus predominately on molar behavior, describing the phenomena under study in terms of nonphysiological, psychological constructs. Biological researchers, in contrast, tend to study the problem at the level of molecular behavior, using neurophysiological constructs. There is often little interaction between these two levels of research, resulting in an absence of a comprehensive theory or explanation of drug-taking behavior. The problem becomes even more complex when it comes to explaining differences in drug-taking among different ethnic groups.

Psychological explanations of drug-taking behavior have particularly failed to account for the inclusion of possible biological or biochemical factors that may contribute to such use; cultural factors also tend to be avoided or minimized. Such explanations thus tend to be unidimensional, rather than multidimensional, and generally focus on a single domain as an explanatory concept to attempt to account for the interrelationship of different contributing factors.

It is clear that any attempt to develop an explanation of drug-taking behavior, specifically inhalant use, needs to pursue how biological factors, which involve genetic and neurochemical processes, and behavioral factors, which include psychological, social and cultural factors, interact. It is also necessary to identify: (a) which factors mediate *against* drug involvement, (b) which factors contribute to the initiation of drug-taking behavior, and (c) which factors reinforce drug-taking behavior. As Cacioppo and Tassinary (1990) noted:

> New conceptual models and research designs for inferring psychological significance from physiological signals may contribute, at least in a small way, to progress toward answering important questions about the relation between complex psychological processes and physiological events, and to a more consistent and productive use of physiological phenomena in the social and psychological sciences. (p. 27)

Drug-taking behavior is a function of the individual (both biologically and psychologically), one's environment (setting), and the drug(s) involved. But such behavior is sometimes more of a function of the individual, or of the setting, or of the availability of and the opportunity to try drugs. What would be helpful is the development of a model that allows these dimensions to be treated both separately and interactively. Such a framework will also make it possible to differentiate among, and study separately and in combination, different factors that represent broad spectrums of behavior. Such factors, in more exact terms, would incorporate three major domains that are known to be involved in use or nonuse of drugs: biological, psychological and sociocultural.

Drawing on the above discussion, a basic and preliminary interactive model is illustrated in Figure 1 to provide a perspective for developing a theory of drug use, with special emphasis on inhalant use. The proposed model obviously cannot account for all the complexities involved in drug use. It is thus an oversimplification of a more highly complex structure, one that can be better defined through subsequent hypothesis testing. In the interim, it serves as a means of presenting an abstraction of some of the elements and processes that influence drug-taking behavior. The model, however, is consistent with some early and contemporary theories of drug-taking behavior (cf., Clayton, & Voss, 1981; Huba & Bentler, 1983; Kaplan et al. 1984; Severson, 1993).

The model is based on the assumption that drug-taking behavior, or the choice to not use drugs, involves the interaction of three major domains: Biological, Psychological and Sociocultural. Each domain plays a different role or exerts a stronger influence at different stages of development and at different stages of drug use; they also interact to modify each other.

The sociocultural dimension represents the *external* elements that influence behavior. Incorporated within this domain are such phenomenon as family and social background, family relationships, values and customs, drinking or drug history, social stability, acculturation stress (as appropriate for different ethnic groups), cultural practices and traditions, environmental related stress, and attitudes toward drug laws, among other events. Also included in this domain are some specific drug-related variables, such as culturally-related drinking or drug-taking practices. This domain specifically interacts with the psychological domain with respect to such variables as availability of drugs, opportunity to try drugs, and special circumstances of initiation.

The influence of acculturation stress has long been an area of concern to researchers (e.g., Ruesch et al. 1948), who have recognized that cultural change is related to crime, suicide and mental health, among many other events. Three possible responses are possible when a person has to adjust to a new culture: (a) establish or maintain ties with one's cultural group in

FIGURE 1. A Biobehavioral Model of Drug-Taking Behavior

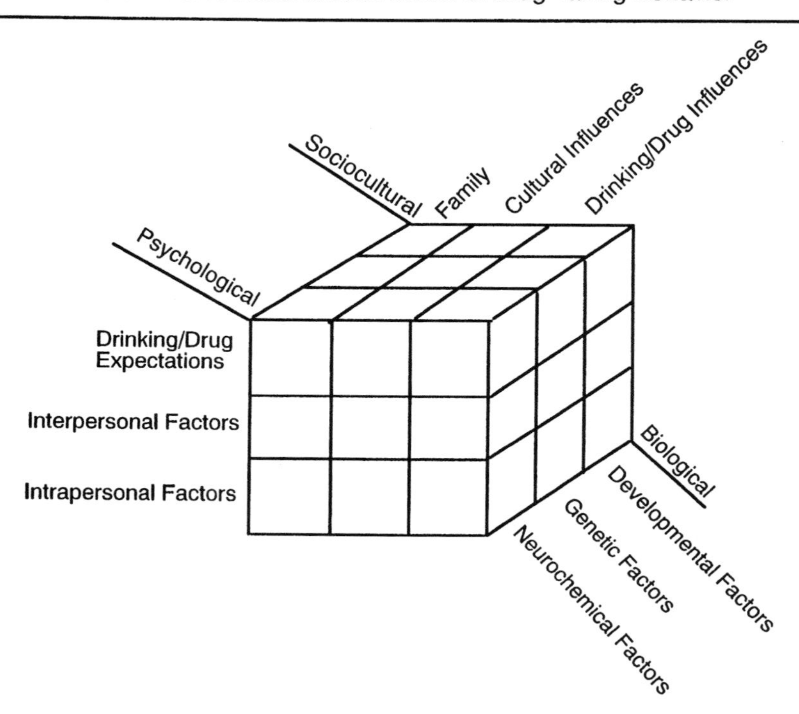


order to reestablish or continue the traditional culture in the new setting; (b) behave as if one is in the traditional culture, despite new surroundings, and let others know as much as possible that they have to adjust to him/her; and (c) accepting the new values and behavioral patterns. Each choice has consequences for the person's physical and mental health, as well as personality structure. Understanding the role that acculturation stress plays in drinking and drug-taking behavior, particularly among affected children, thus becomes instrumental when addressing substance abuse problems in cultural groups undergoing changes in their traditional behaviors, either over a long time period or suddenly and dramatically.

Biological factors delineate the *internal* elements that can influence behavior, represented by genetic, neurochemical and developmental factors or processes. This domain thus represents both biological and developmental processes or factors that contribute to use or nonuse of drugs, and also represents the processes that are involved with the distribution of drugs in the body, the physiochemical mechanisms implicated in a drug's effect, and the metabolism and elimination of drugs from the body.

The psychological domain represent both *inter- and intrapersonal* events. Interpersonal events incorporate the broad range of factors traditionally investigated by psychologists such as personality, needs, traits and tendencies. Intrapersonal variables represent personal, family, community, school and work relationships, as well as how the person perceives and relates to her/his environment. These two concepts have been operationalized in a myriad of psychological research studies.

The psychological domain provides data about what personality factors are strongly related to use or nonuse of drugs. It also, in interaction with the sociocultural domain, yields information about which personality factors are subject to environmental influences that may enhance or inhibit drug-taking behavior. Other types of interactions or influences can also be addressed such as whether and under what conditions support systems (e.g., family) exert greater persuasion than peer influence.

The psychological dimension is critical because it filters and thereby alters the influences of environmental variables. Although individuals will take into consideration the immediate social forces and implications of drug use that impinge on them, and which are embedded in the larger cultural forces that determine identity, role, standards, and conformity, the drug-taking decision is nevertheless a personal assessment of both social and psychological needs (Segal, 1978).

Included in the psychological domain is a drinking/drug expectation factor. This factor is not frequently considered by researchers in their quest to explain drug-taking behavior. It is, nevertheless, an important element in understanding such behavior, one that interrelates closely with the sociocultural domain. The drinking/drug expectation factor represents the nature of the drug(s) experienced, the mode of use, dose, and frequency of use, a person's expectation of and actual behavioral response to a drug, the emotional state at the time the drug is taken, and the setting in which drug taking occurs.

The importance of the user's expectations about a drug's effect is illustrated in the following anecdote regarding an LSD user who was given a placebo, but told that he was taking LSD (cited in Claridge, 1972). Following ingestion of the LSD placebo, the subject reported:

> A lot of strange shapes and brilliant color, after-images, as if I looked through pebble-finished glass, particularly this morning. Especially this morning colors were more brilliant than I have ever experienced. Voices were at times somewhat in the distance along with a feeling of not being in a real situation, a dream kind of state, time is dis-

torted, goes rather slowly, an hour is only 10 or 15 minutes when I look at my watch. (cited in Claridge, 1972, p. 113)

This account "is a perfect description of the LSD state. Interestingly enough, when this subject was told what he had been given and then a week later given a normal dose of LSD, he reported that the drug had no effect at all" (cited in Claridge, 1972, pp. 113-114). Thus the anticipated effect of a specific substance may be so strong that even when placebos are administered to unsuspecting individuals, drug effects are reported–a phenomenon that indicates that psychological factors (set) are strongly involved in the nature and quality of a drug experience.

It is interesting to note that the concept of expectation (set) expressed above, has apparently been incorporated into a conceptual theory, the Theory of Reasoned Action (Fishbein & Ajen, 1975), that is currently being applied to drug research, specifically drug prevention research, and to HIV-risk reduction research. The theory holds that individuals consider the implications of their actions before they decide to act or not. Thus a person's intention to perform or not is the immediate determinant of the action; and two factors contribute to one's decision: (a) attitudes toward the behavior, and (b) the person's perception of the social pressures put on her/him to act or not act, called the subjective norm. Fishbein and Ajen's theory thus provides a means for a more in-depth study of how one's expectations, or intentions, influence drug-taking behavior.

The nature of the person's subjective interpretation of the drug experience is also an important element in determining whether drug use is sustained or not. A person's expectations about a drug's effect, and the nature of the drug experience, also interact with intrapersonal factors, and with both interpersonal and sociocultural factors.

### *Domain Interactions*

That the three domains exert an influence on each other is an inherent assumption of the model. The nature of the specific or causal relationships that exist among the variables needs to be derived from empirical research. Generally, however, from a conceptual viewpoint, it may be argued that the biological factors are fundamental to human behavior and underlie all behavior in terms of possible genetic predisposition. The merit of this argument *is not at issue* in the present model because the primary interest is chiefly in the temporal sequence of human behavior *at a point after which biological factors may have exerted their greatest influence,* and at the stage in which biological components are evidenced in the behavioral realm. Thus, while it is acknowledged that what is called personality may have biological roots, the concern is with those aspects of behavior that exist at an explanatory level beyond direct biological causality.

In all, the proposed scheme represents a complex interaction of sets of variables. It may be helpful to provide a brief example of how the model can be utilized to develop a fundamental theoretical understanding of drug-taking behavior.

A basic assumption is that the individual's perception of the expectation of some form of social and psychological gratification obtained from drug use is a crucial motivating factor involved in setting drug-taking behavior into motion. Once such action is initiated, both internal and external forces interplay to mitigate for or against trying a drug. In this process, elements from the psychological domain, such as the anticipated effects of a drug, might contribute to the decision to either continue to use a drug or to experience another drug, and the context in which the drug is used may contribute, in large part, to whether the drug's effect is experienced as positive or not. Thus, the nature of the drug used, the expectation of its effect(s), and the self-reported phenomenological experience(s) (and whether they are consistent with the known actions of the drug taken), become important factors involved in decisions that affect either experimentation with, continued use, or cessation of drug-taking behavior.

Concerning the individual (psychological domain) and the sociocultural domain, such factors as the relationship between peer group influences (particularly those that may be pitted against family values) and family influences (cf. Kandel, Kessler, & Margulies, 1978), come into play, along with other factors such as availability of drugs, actual opportunity to try a drug, possible fear of arrest, and so on. Such factors need to be examined, singly and in combination with each other, to determine how and to what extent each contributes to the decision to try or not try a drug, and such findings need to be interrelated with the other elements or domains within the model.

It should be noted that a component not explicitly accounted for within this model is that of determining or recognizing stages of drug use. The question of whether such information may be inherent within the model, or a component that needs to be added to the model, should be answered by subsequent research. With respect to the issue of stages of drug use and whether there is a specific order or sequence to drug use, i.e., cumulative or progressive, it is important to note that specific determinants for trying or not trying a drug may not be determinants for continued use of a drug (Segal, 1991). In addition, different selective factors apparently contribute to distinguish between those who decide to try a drug and stop from those who continue to use a drug. After initiation into drug-taking behavior, subsequent behavior can be viewed as a developmental process, involving a series of choices or decisions to stop using, to continue to use, to intensify use or to expand use, that is, to try other drugs. If other drugs are tried, another separate series of decisions take place.

In summary, a preliminary theoretical paradigm has been outlined to help understand the relationship among the many different factors that are related to use or nonuse of drugs. As noted above, the model represents a simplified conceptualization of many different factors that interact to contribute to drug-taking behavior; precisely how these factors ultimately converge to influence the individual with respect to nonuse or to various conditions of drug use is in need of further research.

An advantage of the model is that it provides a frame of reference to study each dimension as an appropriate independent variable and to determine each dimension's relationship to drug use or nonuse alone or in combination with other factors. The model closely parallels the scheme developed by Jessor and Jessor (1980), to conceptualize how human actions interrelate with environmental factors to contribute to drug use, and with Irwin and Millstein's (1986) biopsychosocial perspective on adolescent risk-taking behavior. The strength of such models is that they provide a means of organizing the multiplicity and diversity of events related to drug-taking behavior in such a way that biological, psychological and sociocultural explanations of the data are possible.

## *THEORY APPLICATION*

The proposed theoretical scheme raises the issue of identifying biological, psychological and sociocultural factors that influence the onset and maintenance of drug-taking behavior. Such factors may be referred to as *at-risk* or *risk*-factors. Irwin and Millstein (1986) showed, for example, how biological maturation may affect specific psychosocial changes and the onset of risk-taking behavior, such as substance abuse. The discussion that follows elaborates on how components of the above model may interact, resulting in initiation of drug-taking behavior, and attempts to identify some specific risk-factors that may underlie or contribute to drug-taking behavior.

Prior to the discussion, however, two terms or conditions need to be elucidated–*predisposition* and *precipitating*. Predisposition is defined herein as *the relation obtained when a given phenomenon, which precedes a certain other event in time, either (a) makes the subsequent event occur, (b) makes the subsequent event occur more frequently, or (c) makes the subsequent event occur more intensely.* The predisposing event may be a gene-determined characteristic favoring the development or acquisition of a certain trait or quality, such as a disease, or it may be an environmentally-influenced factor, such as parental drug use.

Precipitating factors, which serve to set behavior into motion, are defined as *a given identifiable phenomena or stimuli that evokes an immediate response that is either (a) highly energized, or (b) highly intense.* Both perception of, and reaction to a precipitating event may be influenced by predisposing factors. That is, earlier (predisposing) experiences contribute to define what the precipitating stimuli is and one's reaction to it.

Figure 2 identifies some predisposing biological, sociocultural contextual and psychological drug-related factors that contribute to place young children at increased risk for drug-taking behavior. These predispositions are linked with some possible precipitating factors that heighten the risk for initiation into drug-taking behavior.

The predisposing factors in Figure 2 also reflect a simplified conceptualization of some of the many different events that interact to contribute to drug-taking behavior. Those listed serve as an immediate representation of some of the many phenomena that can influence use or nonuse of drugs.

What follows is a discussion of how the model may be utilized to account for how one dimension or factor specifically contributes to use of inhalants and other forms of drug-taking behavior. This perspective examines the relationship between cultural change, psychological trauma and drug-taking behavior. Contemporary events in Alaska are used as an illustration of how the model can be applied to examine why drug-taking behavior, especially inhalant abuse among Alaskan Native youth, is exceptionally high in the state.

## *HIGH-RISK FACTORS IN ALASKA*

With its predominantly youthful population and its "last frontier" atmosphere, Alaska is a place where drinking and drug-taking behavior are considerably higher than in the "lower-48" states (Segal, 1990). Linked to Alaska's high prevalence of alcohol- and drug-related problems is the fact that Alaska is still a developing entity, struggling to develop a sense of social identity. This situation is especially manifested in the lives of Alaskan Natives, who are having to deal with the impacts of industrialization and modernization on their traditional cultures. Alaska's rapid development has resulted in an alteration of the Alaskan Natives' cultural values and functions, such as social and economic structures, demography, subsistence, nutrition, spirituality, language and health, among others. As noted in a report by the Alaska Federation of Natives [AFN] (1989):

the pace of economic, social and cultural change in Native villages has been so rapid and the change so profound that many Natives have been overwhelmed by a world not of their making–a world of conflicting values and increasingly limited economic opportunity. For many Natives, the sense of personal, familial and cultural identity that is a prerequisite to healthy and productive life is being lost in a haze of alcohol-induced despair that not infrequently results in violence-perpetrated upon self and family. (pp. 1-2)

FIGURE 2. Some Principal Factors Related to Drug-Taking Behavior

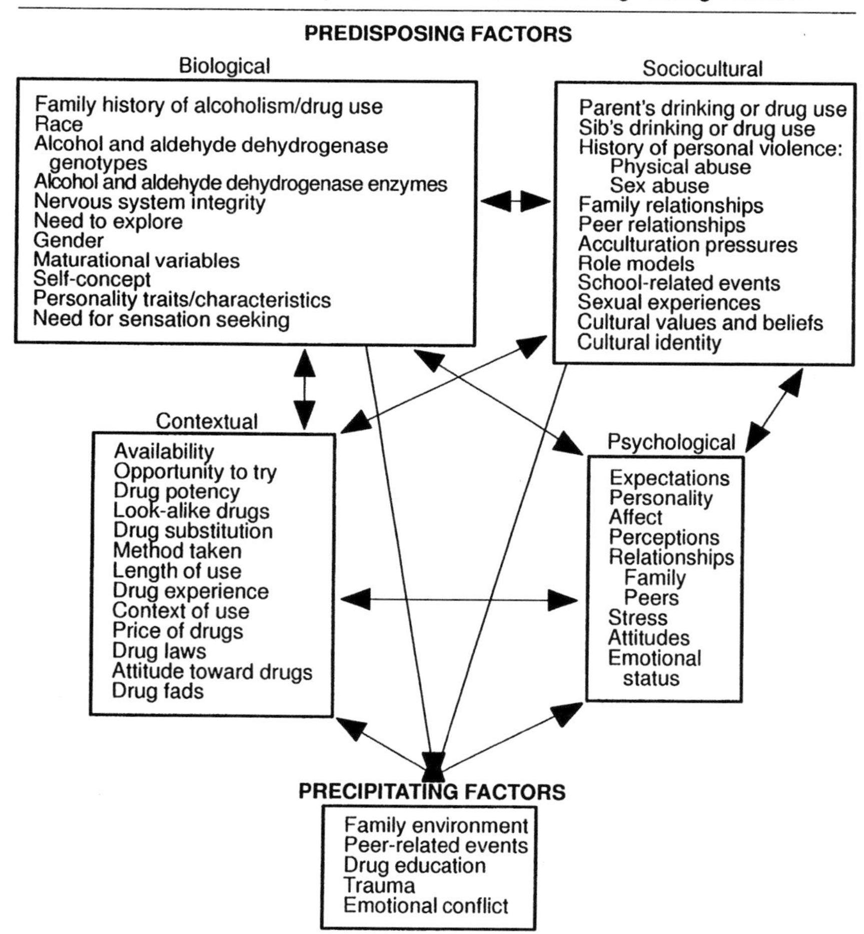

Although despair is not applicable to all Alaskan Native people, those affected, particularly young adults, have created a "plague" of alcohol related behaviors that involves violence and death that disrupts family and community life among Native communities in all of Alaska (AFN, 1989).

Inhalant abuse among Alaskan youth has resulted in similar havoc. The distress within the Alaskan Native community over inhalant abuse came to the forefront following two publicized inhalant abuse deaths. One report (Doto, 1992) described a fatality that involved a 20-year old man, living in a small Native village, who succumbed after sniffing rubber cement with friends behind his parents' house. It was noted that inhalants were popular among the youth in his village, and that several teenagers had also died in other villages in the state.

The second report described the death of a young Native boy from a small village who died from sniffing gasoline. His death was perceived by elders as underscoring the serious inhalant abuse problem among Alaskan Native youth, which they considered to be an "epidemic" (Pagano, 1993). The youth's death elicited the following reactions from village people concerning inhalant abuse: "I see it weekly, in almost every village I visit"; "Gas is everywhere"; ". . . a core group, of three or four young men . . . were enticing kids as young as 7-years old, a first grader, to get involved"; and "I think the least you would find in most villages was 20 percent 'experimentation' rate" (cited in Pagano, 1993, p. B1).

The disruption of the Alaskan Natives' traditional lifestyle and cultural values has also contributed to the development of a host of new problems, including but not limited to new diseases, the effects of accidents and, as indicated above, the effects of new substances (alcohol and other drugs). Berry (1985), noted that many of the cultural changes have health consequences. He stated that such consequences occur

> not simply because {cultural changes} they are newly-introduced . . . but also because the widespread occurrence of cultural loss and social disintegration renders individuals more susceptible to these novel elements; thus, the new concept of *stress* is equally relevant to loss of physical health as to mental health. (p. 22)

One of the byproducts of acculturation is conflict, resulting from pressures exerted by the dominant culture to have the minority group change its way of life. When conflict and stress or tension are present, which are heightened when the minority group resists change and no solution is in sight for the affected group to preserve its ways, the minority culture may be in crisis (Berry, 1985). This crisis is perpetuated over time, depending on the extent of adaptation that takes place, but the stress may nevertheless become cumulative, both on a group and individual level. A culture in crises thus experiences more stress-related problems such as homicide, suicide, family violence, child abuse (both physical and sexual), drinking and other forms of drug-taking behavior, including inhalant abuse, phenomena that are well-documented among Alaskan Natives (Middaugh et al. 1991).

Alaska's development has significantly affected the ability of its Alaskan Natives to maintain their heritage of living with nature in Arctic and subarctic environments, and has resulted in the perpetuation of acculturation stress. Berry (1985) has defined acculturation stress as "stress related to cultural change resulting from continuous, first-hand contact between two distinct cultural groups." Acculturation stress occurs at both a group and individual level, the latter of which is referred to as *psychological acculturation* (Berry, 1985).

Acculturation stress results in cultural disruption that affects normal or traditional behaviors, represented by the sociocultural and psychological domains described above. Additionally, when traditional child rearing values and practices, as well as other aspects of cultural behaviors involved in child rearing are impacted–or even lost to the culture–the result, such as a lack of cultural identity, appears to serve as a particular susceptibility to early involvement in drug-taking behavior. The effects of cultural disruption can thus lead to an *at-risk predisposition* for drinking and other forms of drug-taking behavior early in one's development. These conditions can also affect or involve the biological domain (e.g., Fetal Alcohol Syndrome, or a drug-related birth defect, malnutrition, etc.) or the psychological domain (e.g., physical or sexual abuse).

Cultural factors and the effects of personal violence are well documented in the literature as contributing to drug-taking behavior. Its role has been particularly well formulated in the studies of Oetting and Beauvais (cf. Oetting, Edwards & Beauvais, 1989; Oetting & Beauvais, 1990-91), and Oetting (1993) has recently developed a comprehensive theory describing the relationship between cultural identity and drug-taking behavior. Oetting's theory specifically lends itself to an interpretation of how the domains posited here may interact, with particular insight into how environmental factors are involved in drug-taking behavior.

Another important predisposing factor is child abuse, which is strongly related to acculturation stress. Alaska is an especially violent place for children, with 1,392 reported cases of child abuse for FY 1993, and with over 2,000 unsubstantiated reports for the same time period (Severson, 1993). It has been noted that "families are under much more severe stress than they ever have been and we're seeing a greater depth of violence in families" (Weber, cited in Severson, 1993, p. F3). Child abuse is considered an epidemic within Alaska, with children being hurt at rates higher than the national average. Sixty-three out of every 1,000 children in Alaska were alleged to be victims of child abuse in 1992; the national rate is 39 children per 1,000 (Severson, 1993).

A child's development will be affected by child abuse, in any form. Such abuse, as the child matures, will most likely contribute to a poor self-image, to faulty perceptions of the child's social environment, and to not trusting adults, among other adverse affects. Bryer et al. (1987), for example, found that over three-quarters of their sample ($N = 66$) of psychiatric patients had been physically and/or sexually abused at some time during their lives. Ireland and Widom (1994) also found that early childhood victimization was related to subsequent arrest for alcohol– and/or drug-related offenses. A child's adverse experiences will also influence peer relations, and contribute to the extent to which s/he will identify with peers. Additionally, an early negative childhood experience, particularly sexual abuse, can sufficiently traumatize a child to the extent that its effects become malignant and manifested as a *post-traumatic stress disorder* (PTSD). Herman (1992) noted that "with severe enough traumatic exposure, no person is immune" (p. 57), and the condition [PTSD] can be experienced in the immediate aftermath of the event or in years subsequent to it. The essential element of child abuse is the physical and psychological impact it has on the child, and involvement with drugs or drinking in later life can serve as a means of helping one to escape from unwanted thoughts and emotions.

PTSD is described by the American Psychiatric Association (1994) as an anxiety disorder characterized by the re-experiencing of a traumatic event, accompanied by symptoms of increased arousal and by avoidance of stimuli associated with the trauma. Zweben et al. (1994) indicated that "the stressor or trauma need not be directly experienced, but may instead be witnessed (such as in children exposed to urban violence); the fear of the trauma is a sufficient precursor to the development of PTSD" (p. 329). The relationship between trauma, specifically sexual assault, PTSD and substance abuse is well documented (cf. Brown, 1994; Harvey et al. 1994; and Zweben et al. 1994). Indeed, as Harvey et al. stated: "Clinical reports suggest that unresolved trauma-related symptoms can contribute to relapse, as individuals may eventually return to alcohol and other drug use to cope with unresolved long-term effects of trauma" (p. 361). Zweben et al. (1994) also stated that: "Clinicians have noted that many patients describe a pattern of alcohol or drug use motivated by a desire to obscure or escape from uncomfortable experiences. Some are now conceptualizing the alcohol and other drug use as one *method* (among several) to dissociate painful states" (p. 330).

Dick, Manson, and Beals (1993) reported that American Indian youth "may turn to alcohol to relieve symptoms of negative affect" (p. 175), and that the youth in their study "showed an association between stressful life events and quantity/frequency of alcohol use" (p. 176). Khantzian (1985) described such drug-taking behavior as consistent with a "self-medication hypothesis." Self-medication refers to use of drugs to medicate oneself for a range of psychiatric problems and painful emotional states. The hypothesis infers that the user selects drugs to achieve a "short-term effect to help cope with distressful subjective states, and an external reality otherwise experienced as unmanageable or overwhelming" (Khantzian, 1985, p. 1263).

A recent review of inhalant abuse research and a longitudinal study of drug abusers led to the conclusion that there is a relationship between inhalant abuse and psychiatric illnesses (Compton et al. 1994). Specifically, these researchers found an association between inhalant abuse and antisocial personality disorder, social phobia, alcohol dependence, tobacco dependence, injection drug use, and use of amphetamines, sedatives, cocaine, opiates, PCP, and hallucinogens. Inhalant abuse was specifically described by Compton et al. as a precursor to other substance use and to psychiatric illness, a finding consistent with an earlier report by Dinwiddie and Cloninger (1991). These researchers concluded that solvent use was associated with other forms of drug use and with a diagnosis of antisocial personality disorder. These findings are consistent with Oetting and Webb's (1992) report that inhalant users are among the most dysfunctional of all drug users; they also suggested that inhalant abusers may have been subjected to some form of trauma in their development to make them so vulnerable to such emotional distress subsequent to initiation to inhalant use. They also supported the hypothesis that inhalant use may be a form of self-medication, serving to mask a sense of personal distress.

Acculturation stress and/or personal violence experienced early in life may thus predispose one for drug-taking behavior. This predisposition occurs because one may tend to develop a unique perception of risk-taking behavior. Risk-taking behavior is defined herein as *choosing to or not to do something in which the outcome presents a threat to one's integrity, either physically, psychologically or both.* Choosing to inhale a substance, such as gasoline, places the person at risk, while not choosing to use a seat belt, or to "rope up" while traversing a glacier, also places one at risk.

The key concept involved in this definition is *choice,* which infers that the perception of the risk involved is subjective (Carney, 1971). The ability to make a subjective choice with respect to determining its consequences is age-related, that is, as one increases in age, one develops a greater repertoire of experiences on which to base one's choice. For example, an adult choosing to smoke makes this choice despite the fact that smoking has been shown to threaten health. In contrast, a child who starts smoking may not know enough about the adverse effects or be concerned with health problems to consider smoking a risk.

Thus the early use of inhalants and other drugs for many youth, particularly for Alaskan Native youth, may represent behavior that is precipitated by immediate events, but the effects of the inhalant-related experience may in turn prove rewarding because it alleviates a sense of personal distress. Heavy reliance and continuous use of inhalants, or the emergence of other forms of substance use, can contribute to problems of abuse and dependency, but the self-medication effects tend to be more compelling than any adverse health, social, behavioral or legal consequences. Violent, prolonged, or intrusive experiences at an early age, including the effects of acculturation stress, represent events that can result in a traumatic experience that is beyond the adaptive capacity of the child, resulting in a long-lasting traumatic syndrome which, at a later age, may be self-medicated through drinking and other forms of drug-taking behavior. For some, the more severe the trauma, the earlier they may initiate inhalant abuse due to a greater vulnerability to precipitating events.

## *THEORY TESTING*

The above discussion presented a theoretical review concerning initiation into inhalants, and other substances, based on a theorized relationship between personal violence, acculturation stress, post-traumatic stress disorder and drug-taking behavior. Although this relationship has previously been well conceptualized (cf., Zweben et al. 1994), it remains in need of empirical testing to determine if it has sufficient strength to generate implications which exist beyond the initially formulated propositions. An examination of these implications may support the inferred relationship, or may generate alternate explanations. Figure 3 presents two models that are in need of testing to determine the relationship between early adverse experiences, PTSD and drinking and drug-taking behavior, principally inhalant abuse. These models are based on the interactions between the psychological and sociocultural domains. While the biological domain contributes to drinking and drug-taking behavior, its influence is not accounted for in this early conceptualization stage.

Model 1 (in Figure 3) suggests that personal violence and acculturation stress are independent of each other, but that either can contribute to PTSD, which in turn is related to drinking and drug use. Model 2 suggests

FIGURE 3. Theoretical Models of the Relationship Between Personal Violence, Acculturation Stress, and Drinking and Drug-Taking Behavior

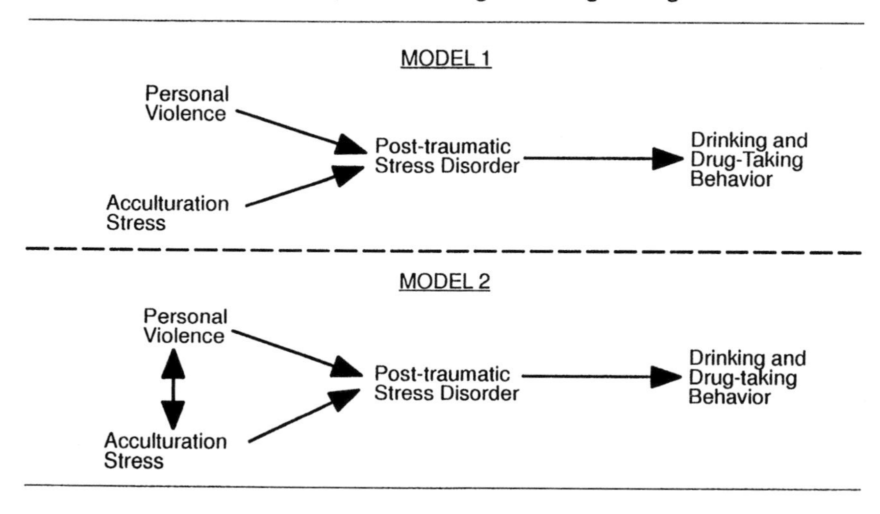


that personal violence and acculturation stress are interdependent, and that they are both related to the onset of PTSD and subsequent drinking and drug use.

Having proposed these two causative models, the task becomes one of empirically testing their efficacy. Further research is required to determine which is more predictive of inhalant abuse, drinking and other forms of drug-taking behavior.

## *CONCLUSION*

Researchers, in order to seek explanations of drug-taking behavior, have advanced a number of general theories that have helped to shed light on many causal factors, but such theories have not fully explored the relationship between behavioral or psychological, sociocultural and biological factors and vulnerability to drug-taking behavior.

What has been proposed herein is a framework for developing a model that attempts to identify how some specific factors are related to drug-taking behavior. These special factors are acculturation stress, the early experience of personal violence, and the onset of post-traumatic stress disorder. It has been suggested that acculturation stress and/or the experience of personal violence are predisposing factors that contribute to the onset of a post-traumatic stress disorder, which in turn serves as a precipitating factor (motivator) related to initiation into drug-taking behavior. In this context, drinking or drug use represents a self-medication process that is linked to motive for use. The heuristic value of this scheme is in need of testing, and two models that can be tested empirically have been advanced.

The questions raised earlier in the paper about why youth use inhalants can begin to be answered on the basis of the model proposed here. The first two questions were, "What attracts youth to inhalants?" and "Why do youth put themselves at risk by using inhalant substances?" Inhalant abuse, for many youth, appears to revolve around use of a selected substance to obtain what might be characterized as a "cheap high." Inhalants, which are inexpensive and available to minors in various forms (such as gasoline), produce euphoria and alters consciousness, experiences that youth may be seeking that, in addition to emulating what older people do, may also represent a form of risk-taking behavior. While this risk-taking behavior, for some youth, may be related to the effects of acculturation stress or some form of a violence-related trauma experienced earlier in life, its use nevertheless represents an immediate means of coping with personal discomfort (PTSD). It is thus possible that the prevention messages urging caution because of the dangers inhalants present, may make inhalants attractive for youth predisposed to risk-taking behavior. Others may find it a quick and easy way to escape from reality. In either case, the desired results appear to be sufficiently rewarding to the user so that inhalant use may be repeated.

Another question was, "Why are some youth, especially American Indian and Alaskan Native youth, more vulnerable to using inhalants than youth from other racial groups?" The answer to this question is that both groups tend to experience the effects of what Oetting (1993) has described as "cultural failure," which involves a lack of identification with traditional cultural values and practices. This absence leads youth to engage in risk-taking behaviors early in their lives to fill needs that cannot be achieved through culturally approved practices. Inhalants, as discussed above, are a readily available means to "act out." Acculturation stress is particularly intense in isolated communities in Alaska or on reservations where people live under difficult health and social conditions. Inhalants, or other forms of substance abuse, may thus be a manifestation, albeit a self-destructive one, of acculturation stress or child abuse, which is now recognized to occur in males as well as females (Dunn et al. 1994; Harvey et al. 1994). Zweben et al. (1994) noted that:

> Substance abuse is only one of many sequelae of child-hood abuse, including low self-esteem, depression, suicide, anxiety, difficulties in interpersonal relationships, sexual dysfunction, and a tendency toward revictimization. . . . girls are at approximately twice the risk as boys, with no increase in adolescence . . . these children are at nearly a fourfold increased lifetime risk for any psychiatric disorder, and a threefold risk for substance abuse. . . . It appears that severe childhood incest may result in the adult survivor being unable to reach the socioeconomic level of . . . parents. . . . [It has been] suggested that the survivor[s] of severe trauma . . . absorbed . . . [their] energies coping with the profound consequences of the abuse, while . . . [their] peers, less constrained, were better able to actualize their potential. . . . What appears as lack of client motivation may be one of many reverberations of sexual abuse in childhood, and this awareness needs to be integrated into treatment efforts, especially with the economically disadvantaged. (p. 331)

It is thus apparent that trauma contributes to a cluster of symptoms, best represented by PTSD, which is directly linked to substance abuse. Herman (1992) indicated that if treatment for substance abuse is to succeed, it must address underlying problems, such as PTSD, as well as the substance-abusing behavior.

The postulation of two special causative factors related to initiation into drug-taking behavior has clear implications for prevention and for clinical intervention. Concerning prevention, rather than developing education/prevention programs that primarily focus on the adverse effects of drugs, efforts would need to be directed at reducing circumstances that contribute to an increased risk for initiation into drug-taking behavior, such as helping impacted communities deal with acculturation stress. Effort would also need to be directed at helping youth recognize and begin to deal with their own personal distress, rather than seeking escape through drug-taking behavior. These efforts are especially necessary when addressing minority populations such as American Indians and Alaskan Natives, who continue to experience the process of acculturation.

Although acculturation cannot be stopped (or reversed), ways have to be developed to modify acculturation pressures so that change becomes opportunities rather than problems for the impacted group (World Health Organization [WHO], 1985). Within this context, many of the problems specific to the dominated group need to be addressed, such as drinking and drug use, violence, family problems, and so on. A distinction, however, needs to be made between long-term solutions (which emphasize prevention) and short-term accomplishments (which emphasize resolving immediate problems). An important element in this process is *control.* Control involves being in charge of one's destiny, that is, having a choice, both at an individual and group level. At an individual level choice means having a sense of control over one's own life and one's immediate family; group control means having a collective right to decide how to live under new conditions (WHO, 1985). Strategies need to be developed to assist minority groups, such as Native people or new immigrant groups clustered in ghetto communities, establish a sense of control and to maintain some aspects of their original social patterns as they function within the dominant culture. The problems related to drinking and drug-taking behavior among dominated cultural groups cannot be reduced unless the underlying contributing factors are addressed, such as the effects of acculturation stress.

There are three major clinical implications derived from the theoretical framework discussed above. The first is the necessity to conduct a comprehensive diagnostic assessment with people seeking treatment that includes an environmental history for both childhood and adult life. Second, it is essential to evaluate for PTSD. Thirdly, the assessment procedure has to be culturally relevant when used with members of minority groups.

Although the focus of the discussion has been on seeking an explanation of drug-taking behavior among minority or dominated groups within the larger society, the principles involved pertain to the dominant group as well. Although acculturation stress may not be a factor, other life stresses can be substituted, and the dominate culture is not immune to personal violence and its effects. It thus appears that further research is also required to begin to explore more fully the relationship between trauma, PTSD and drug-taking behavior among the general population.

## REFERENCES

The Alaskan Federation of Natives (1989). *The AFN report on the status of Alaska Natives: A call for action.* Anchorage, AK: AFN.

American Psychiatric Association. (1994). *Diagnostic and Statistical Manual of Mental Disorders (4th Ed.).* Washington, D.C.: American Psychiatric Association.

Agarwal, D. H., and Goedde, W.H. (1989). Human Dehydrogenases: Their role in alcoholism. *Alcohol, 6,* 517-523.

Beauvais, F. (1992). Volatile solvents abuse: Trends and patterns. In C. Wm. Sharp, F. Beauvais, & R. Spence (Eds.), *Inhalant abuse: A volatile research agenda* (pp. 13-42). (NIDA Research Monograph 129.) Rockville, MD: National Institute on Drug Abuse.

Beauvais, F., & Segal, B. (1992). Drug use patterns among American Indian and

Alaskan Native youth: Special rural populations. *Drugs & Society, 7*(1/2), 77-94.

Berry, J. W. (1990). Acculturation among circumpolar peoples: Implications for health status. *Arctic Medical Research, 40*(21), 21-27.

Brecher, E. M. (1972). *Licit and illicit drugs.* Boston: Little Brown & Co.

Brown, S. (1994). Alcoholism and trauma: A theoretical overview and comparison. *Journal of Psychoactive Drugs, 26*(4), 345-355.

Bryer, J. F., Nelson, B. A., Miller, J. B., & Krol, P. (1987). Childhood sexual and physical abuse as factors in adult psychiatric illness. *American Journal of Psychiatry, 144*(11), 1426-1430.

Cacioppo, J. T., & Tassinary, L. G. (1990). Inferring psychological significance from physiological signals. *American Psychologist, 45*(1), 16-28.

Caputo, R. A. (1993). Volatile substance misuse in children and youth: A consideration of theories. *The International Journal of the Addictions, 28*(10), 1015-1032.

Carney, R. E. (1971). *Risk-taking behavior. Concepts, methods, and applications to smoking and drug abuse.* Springfield, IL: Charles C. Thomas.

Claridge, G. (1972). *Drugs and human behavior.* Middlesex, England: Pelican.

Clayton, R., & Voss, H. (1981). *Young men and drugs in Manhatten: Causal analysis* (Research Monograph No. 39). Rockville, MD: National Institute on Drug Abuse.

Compton, W. M., Cottler, L. B., Dinwiddie, S. H., Spitznagel, E. L., Mager, B. S., & Asmus, G. (1994). Inhalant use. Characteristics and predictors. *The American Journal of Addictions, 3*(3), 263-272.

DeLaRosa, Mario, R. & Adrados, J-L. R. (1993). *Drug abuse among minority youth: Advances in research and methodology.* (NIDA Research Monograph 130). Rockville, MD: National Institute on Drug Abuse.

Dick, R. W., Manson, S. M., & Beals, J. (1993). Alcohol use among male and female Native American adolescents: Patterns and correlates of student drinking in a boarding school. *Journal of Studies on Alcohol, 54,* 172-177.

Dinwiddle, S. H., & Cloninger, R. C. (1991). The relationship of solvent use to other substance use. *The American Journal of Drug and Alcohol Abuse, 17*(2), 173-186.

Doto, P, (Oct., 1992). Stebbins man sniffs glue, dies. *Anchorage Daily News,* October 4, 1992, p. B1.

Dunn, G. E., Ryan, J. J., & Dunn, C. E. (1994). Trauma symptoms in substance abusers with and without histories of childhood abuse. *Journal of Psychoactive Drugs, 26*(4), 357-360.

Fishbein, M. & Azjen, I. (1975). *Belief, attitude, intention and behavior: An introduction to theory and research.* Reading, MA: Addiison-Wesley.

Harvey, E., Rawson, R. A., & Obert, J. L. (1994). History of sexual assault and the treatment of substance abuse disorders. *Journal of Psychoactive Drugs, 26*(4), 361-367.

Herman, J. (1992). *Trauma and Recovery.* NY: Basic Books.

Huba, G. J., & Bentler, P. M. (1983). Tests of a drug use causal model using asymptotically distribution free methods. *Journal of Drug Education, 13*(1), 3-14.

Ireland, T., & Widom, C. S. (1994). Childhood victimization and risk for alcohol and drug arrests. *The International Journal of the Addictions, 29*(2), 235-274.

Irwin, C. E., & Millstein, S. G. (1986). Biopsychosocial correlates of risk-taking behaviors during adolescence. *Journal of Adolescent Health Care, 7,* 82S-96S.

Jessor, R. J., & Jessor, S. (1980). A social-psychological framework for studying drug use. In D. J. Lettieri, M. Sayers, & H. W. Pearson (Eds.), *Theories on drug abuse: Selected contemporary perspectives* (NIDA Research Monograph 30), pp. 102-109. Rockville, MD: National Institute on Drug Abuse.

Johnston, L. D., O'Malley, P. O., & Bachman, J. G. (1993). *National survey results on drug use from monitoring the future study, 1975-1992. Vol 1. Secondary School Students.* Rockville, MD: National Institute on Drug Abuse.

Kandel, D. B., Kessler, R. C., & Margulies, R. Z. (1978). Antecedents of adolescent initiation into stages of drug use: A developmental analysis. In D. B. Kandel (Ed.), *Longitudinal research on drug use: Empirical findings and methodological issues* (pp. 73-100). Washington, D.C.: Hemisphere.

Kaplan, H. B., Martin, S. S., & Robbins, C. (1984). Pathways to adolescent drug use: Self-derrogation, peer influence, weakening of social controls, and early substance use. *Journal of Health and Social Behavior, 25,* 270-289.

Khantzian, E. J. (1985). The self-medication hypothesis of addictive disorders: Focus on heroin and cocaine dependence. *American Journal of Psychiatry, 142*(11), 1259-1264.

Middaugh, J. P., Miller, J., Dunaway, C. E., Jenkerson, S.A., Kelly, T., Ingle, D., Perham, K., Fridley, D., Hlady, W. G., & Hendrickson, V. (1991). *Causes of death in Alaska 1950, 1980-1989.* Juneau, AK: Department of Health & Social Services.

Nathan, P. E. (1990). Integration of biological and psychosocial research on alcoholism. *Alcohol, 14*(3), 368-374.

Oetting, E. R. (1993). Orthogonal cultural identification: Theoretical links between cultural identification and substance use. In M. DeLaRosa & J-L R. Adrados (Eds.), *Drug abuse among minority youth: Advances in research methodology* (pp. 32-56) (NIDA Research Monograph #130). Rockville, MD: National Institute on Drug Abuse.

Oetting, E. R., & Beauvais, F. (1990-91). Orthogonal cultural identification theory: The cultural identification of minority adolescents. *The International Journal of the Addictions, 25*(5A & 6A), 655-685.

Oetting, E. R., Edwards, R. W., & Beauvais, F. (1989). Drugs and Native-American youth. *Drugs & Society, 3*(1/2), 5-38.

Oetting, E. R., & Webb, J. (1992). Psychological characteristics and their links with inhalant use: A research agenda. In C. Sharp, F. Beauvais, and R. Spence (Eds.), (pp. 59-98) (NIDA Research Monograph Series #129) *Inhalant abuse: A volatile research agenda.* Rockville, MD: National Institute on Drug Abuse.

Pagano, R. (May, 1993). Teen deaths raise concern over inhalants. *Anchorage Daily News,* May 17, 1993, pp. A1, A8.

Ruesch, J., Jacobsen, A., & Loeb, M. B. (1948). Acculturation and illness. *Psychological Monographs, 62*(5) 1-40.

Segal, B. (1978). Sensation seeking and drug use. In D. J. Lettieri (Ed.), *When coping strategies fail: Drugs, alcohol and suicide* (pp. 149-166). (Vol. 2, Sage Annual Review of Drug and Alcohol Abuse.) Beverly Hills: Sage Publications.

Segal, B. (1988). *Drugs and behavior: Cause, treatment, and prevention.* New York: Gardner Press.

Segal, B. (1990). *Drug-taking behavior among school-aged youth: The Alaskan experience and comparisons with lower-48 states.* NY: The Haworth Press, Inc.

Segal, B. (1991). Adolescent initiation into drug-taking behavior: Comparisons over a five-year interval. *The International Journal of the Addictions, 26*(3), 267-279.

Segal, B., Duffy, L., Avksentyuk, A., & Thomasson, H. R. (October, 1993). *Alcohol and aldehyde genotypes among Alaskan and Siberian Natives.* Paper presented at the annual meeting of the American Public Health Association, San Francisco, CA.

Severson, H. (1993). A psychometric study of adolescent risk perception. *Journal of Adolescence, 16*(2), 153-168).

Severson, K. (Oct., 1993). Alaska's children are suffering from an epidemic of family violence. *Anchorage Daily News,* Oct. 8, pp. F1, F3.

University of Michigan (Jan., 1994). *Monitoring the future.* News Release. Ann Arbor Michigan, The University of Michigan.

World Health Organization (1985). Problems of family health in circumpolar regions. *Arctic Medical Research, 40,* 7-20.

Zweben, J. E., Clark, H. W., & Smith, D. E. (1994). Traumatic experiences and substance abuse: Mapping the territory. *Journal of Psychoactive Drugs, 26*(4), 327-344.

# Research Topics for the Problem of Volatile Solvent Abuse

Fred Beauvais, PhD

**SUMMARY.** Solvent abuse has received comparatively little attention from the public and from drug abuse researchers and practitioners. This neglect has resulted in a dearth of knowledge regarding this hazardous behavior. Based on the experience of the staff of the Tri-Ethnic Center for Prevention Research, a research agenda is proposed that addresses the major dimensions of solvent abuse. Topics for both prevention and treatment are included. *[Article copies available for a fee from The Haworth Document Delivery Service: 1-800-342-9678. E-mail address: getinfo@haworth.com]*

The Tri-Ethnic Center for Prevention Research at Colorado State University has significant experience with the problem of volatile solvent abuse. Interest in this topic began a number of years ago when it became apparent that one of the populations under research by the Center, American Indian youth, had inordinately high rates of solvent abuse. Of necessity, the Center staff began to gather literature and research data on this topic in order to understand what was occurring among these youth. In subsequent years the staff became very familiar with this area of research and developed a general expertise on solvent abuse. This led to presentations at national and international meetings, the writing of numerous articles, training workshops on prevention and treatment, and consultation with numerous community and professional groups. The latter has in-

---

Fred Beauvais is Senior Research Scientist, Tri-Ethnic Center for Prevention Research, Department of Psychology, Colorado State University.

[Haworth co-indexing entry note]: "Research Topics for the Problem of Volatile Solvent Abuse." Beauvais, Fred. Co-published simultaneously in *Drugs & Society* (The Haworth Press, Inc.) Vol. 10, No. 1/2, 1997, pp. 103-107; and: *Sociocultural Perspectives on Volatile Solvent Use* (ed: Fred Beauvais, and Joseph E. Trimble) Harrington Park Press, an imprint of The Haworth Press, Inc., 1997, pp. 103-107. Single or multiple copies of this article are available for a fee from The Haworth Document Delivery Service [1-800-342-9678, 9:00 a.m. - 5:00 p.m. (EST). E-mail address: getinfo@haworth.com].

 

cluded a close working relationship with the only three solvent treatment programs in operation in North America.

Out of this collection of experiences the Center staff identified a number of important research questions that need answering if we are to be effective in furthering our efforts in both the prevention and treatment of solvent abuse. Many of these questions are not addressed in the current literature; rather they come from staff observations and the observations of those who are in daily contact with the solvent abuse problem. Many of the questions included under the heading of prevention refer to epidemiological inquiries because it is only with a sound knowledge of rates of use and demographic patterns that competent prevention approaches can be designed.

## *PREVENTION QUESTIONS*

*What are the "cultural" meanings of solvent abuse among those who are using?* Bob Trotter, in this volume, addressed this issue among one particular ethnic minority population, but this work needs to be extended more broadly. Because solvent use displays a number of anomalies when compared to other drugs (see Oetting and Webb, 1992 and Beauvais, 1992), it would be important to understand what users are actually thinking when they use or contemplate use of solvents. In our work we have often heard young people say, "No, I don't use drugs, but I do sniff once in a while." What are the distinctions that are being made between solvents and other drugs? Are these perceptions the reason why solvent abuse was increasing through the 80's and early 90's when the use of other drugs was decreasing? Are drug prevention programs excluding solvent abuse as a topic? In contrast to most other drugs, solvent abuse peaks at an early age (about 13); what are the perceptions of these young people regarding solvents? The gender gap is rapidly closing on the prevalence rates of solvent use; do male and females perceive the use of solvents differently?

*Why does the use of solvents often occur in rapidly cycling epidemics?* The episodic nature of solvent abuse in a population is in contrast to other drugs where increases or decreases occur over several year's time. One of the concerns about these epidemics is that while most youth will stop using once the epidemic is over, each wave of use leaves behind a group of vulnerable individuals who will continue use and go on to heavier use. What are the environmental triggers for solvent use? What are the inter- and intrapersonal dynamics that lead to the rapid rise once an epidemic has started? What are the factors influencing the rapid decline of solvent use?

*What are the sociocultural or other conditions that give rise to adult patterns of solvent abuse?* Most solvent abuse occurs among younger people, and occasionally this pattern continues into adulthood. In some locations, however, significant populations of adult solvent abusers have been identified; little information is available regarding the conditions that are associated with this phenomena. One speculation is that it is simply an economic phenomena; a $1.00 can of spray paint, for instance, can keep a person high for a day or more. However, there are many economically depressed communities that do not have an adult solvent abuse problem. Why does solvent abuse become prominent among adults in one community and not another with similar socioeconomic circumstances?

*What are the psychosocial factors associated with solvent abuse among both youth and adults?* It has been established that solvent abusers present a much more dysfunctional psycho-social profile (Oetting and Webb, 1992), but little is known why these factors lead more often to the abuse of solvents. What is the "natural history" of solvent abuse? Is there a common course or progression of use from solvents to other substances or other substances to solvents? Is it the total number of problems experienced by individuals that lead to solvent abuse or are there particular types and combinations of problems? For instance, a common observation of clinicians who treat solvent abusers is that there appears to be a history of very intense sexual abuse. Is this a primary factor or only one of a number of factors?

*What are the most effective prevention messages to counter solvent abuse? Are there age-specific factors that need to be considered?* Because solvent abuse often starts at very young ages there is always the possibility that new, inhalable substances will be suggested to young people in the course of prevention efforts. Solvent containing products are ubiquitous and cannot be eliminated from the environment so care must be taken not to introduce new ideas during this curious and impressionable age. At what age is it appropriate to talk openly about products that can be inhaled for psychoactive purposes? More generally, what are the elements of an effective solvent abuse prevention curriculum?

*Can effective policy measures be introduced that will help limit the availability of inhalable solvents?* In contrast to most other drugs of abuse, solvents have many legitimate uses. Are there ways that producers and sellers of these products can restrict the appeal and access to these products? Can legal means of curtailment be effectively implemented? What has been the impact on the levels of solvent abuse in locations where solvent intoxication has been made a crime?

## *TREATMENT QUESTIONS*

*What is the optimal length of time needed for effective treatment of solvent abuse and what are the stages of treatment?* Solvents are lipophilic and remain in the body for extended periods of time. This would suggest that there is need for a longer detox period than is found for treatment for abuse of other drugs. What length of time is needed? Many of the existing treatment regimes in the drug treatment field call for strategies that require a fairly high level of cognitive functioning. If these strategies are implemented, solvent abusing patients may not be capable of responding, leading to frustration on the part of both patients and treatment staff. Can stages of recovery be identified that can be matched to increasing levels of complexity in treatment activities?

*What is the level and type of neurological damage caused by solvents, and is this damage reversible?* A common belief among both lay people and drug abuse professionals is that solvent abusers incur immediate and irreversible neurological damage from the first onset of use. This belief has given rise to an attitude that treatment time may be wasted on solvent abusers and that efforts should be directed away from them towards abusers of other drugs. Reports from clinicians, however, indicate that a great deal of neurological functioning can be regained. What are the conditions that will facilitate this? Is supportive care only sufficient for this recuperative process? Are there specific activities or treatment modalities that will encourage neural repair or training around existing damage?

*Given the level and breadth of dysfunction found among most solvent abusers, what is the range of treatment interventions that are required for long term sobriety?* It appears that most of the social systems of solvent abusers are maladaptive. Effective treatment would thus seem to have to include restructuring of the family, peer, spiritual, school, and work relationships. Can adequate resources be brought to bear on this constellation of problem areas? What are the minimum interventions needed for maintaining sobriety?

*Are there unique issues that must be addressed in relapse prevention and aftercare plans for solvent abusers?* Many solvent abusers report, for instance, an intense craving that is triggered by certain tastes and smells. How can the patient be prepared for this? Many solvent abusers, primarily adults, alternate the use of solvents with alcohol and report that alcohol is one of the triggers for relapse. Reinforcing this pattern is the attempt by solvent abusers to sustain their level of intoxication from alcohol by switching to the much cheaper and available solvents when their money runs short. What are the elements of alcohol abuse treatment that need to be incorporated into solvent abuse treatment programs?

*Can solvent abusers be treated effectively in a general drug treatment program?* Many treatment programs report that solvent abusers do not do well when in treatment with abusers of other drugs. It could well be that solvent abusers differ in too many ways and that their needs are incompatible with existing treatment programs.

*Can chronic solvent abuse be effectively treated in an outpatient setting?* Once again the breadth and severity of problems experienced by this population may preclude normal outpatient approaches. For instance, one of the known characteristics of solvent abusers is their propensity to avoid getting into treatment in the first place and then leaving treatment before any therapeutic gains can be made. Can more rigorous and controlled outpatient procedures be developed that will insure treatment compliance?

*Is there a fetal volatile solvent syndrome?* One of the most commonly asked questions from those working with solvent involved populations is whether or not fetal damage occurs among pregnant women who use solvents. This question undoubtedly arises from the current interest in FAS, and from the observation that infants of solvent abusing mothers appear to have a number of developmental problems. Attributing these problems to fetal solvent exposure is problematic for a number of reasons. First, pregnant women who use solvents are likely using other substances, most commonly alcohol, and sorting out the effects of the different chemicals is difficult. Added to this are the differential effects of multiple solvents that may have been used during pregnancy. Further complicating the picture is the likely neglect that a developing fetus or infant are exposed to. Solvent using mothers are a highly dysfunctional group in a number of ways, and the physical and emotional needs of their children are commonly overlooked. It would be useful to study the behavior of solvent using women during pregnancy in order to gain insight into the potential teratogenic effects of solvents, and to also provide definitive information for treatment of this high risk group of women and their children.

The above research agenda is ambitious. It calls for an awareness of the extent of solvent abuse and the resources with which to address it. Unfortunately, solvent abuse has a very low profile within the substance abuse field, and the funding for research is scarce along with very few researchers with an interest in the topic. Only with greater public concern and professional effort will we come to an understanding of the causes and consequences of this highly dangerous behavior.

## REFERENCES

Oetting, E. R., & Webb, J. (1992). Psychosocial characteristics and their links with inhalants: A research agenda. In C. Sharp, F. Beauvais, & R. Spence (Eds.), *Inhalant Abuse: A Volatile Research Agenda,* (NIDA Research Monograph No. 129). Rockville, MD: National Institute on Drug Abuse.

Beauvais, F. (1992). Volatile solvent abuse: Trends and patterns. In C. Sharp, F. Beauvais, & R. Spence (Eds.), *Inhalant abuse: A volatile research agenda,* (NIDA Research Monograph No. 129). Rockville, MD: National Institute on Drug Abuse.

# Index

Page numbers followed by "t" indicate tables; page numbers followed by "f" indicate figures.

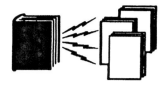

***COSTS:*** Please figure your cost to order quality copies of an article.

1. Set-up charge per article: $8.00
   ($8.00 × number of separate articles) __________
2. Photocopying charge for each article:
   1-10 pages: $1.00 __________
   11-19 pages: $3.00 __________
   20-29 pages: $5.00 __________
   30+ pages: $2.00/10 pages __________
3. Flexicover (optional): $2.00/article __________
4. Postage & Handling: US: $1.00 for the first article/
   $.50 each additional article __________
   Federal Express: $25.00 __________
   Outside US: $2.00 for first article/
   $.50 each additional article __________
5. Same-day FAX service: $.35 per page __________

**GRAND TOTAL:** ____________

---

***METHOD OF PAYMENT:*** (please check one)

❑ Check enclosed ❑ Please ship and bill. PO # ________________

(sorry we can ship and bill to bookstores only! All others must pre-pay)

❑ Charge to my credit card: ❑ Visa; ❑ MasterCard; ❑ Discover; ❑ American Express;

Account Number: ________________ Expiration date: __________

Signature: ✗ ________________________

Name: ________________ Institution: ________________

Address: ________________________________

________________________________

City: ________________ State: ________ Zip: ________

Phone Number: ________________ FAX Number: ________________

---

**MAIL or *FAX* THIS ENTIRE ORDER FORM TO:**

Haworth Document Delivery Service
The Haworth Press, Inc.
10 Alice Street
Binghamton, NY 13904-1580

**or FAX:** 1-800-895-0582
**or CALL:** 1-800-342-9678
9am-5pm EST)